'Thought provoking … [Paul Bogard] is convincing that artificial light has blinded us to the beauties of the night. Perhaps it's time to follow him over to the dark side'

' on Sunday

'Bogard sets about his investigations with an energetic purposiveness and enterprise ... To seek to let back in a little of the lost starlight and allow more of nature's shadow to reassert its balm seem to me both modest and wholesome aims, and Bogard's book does much to make a case for them' SALLEY VICKERS, *Observer*

'Bogard takes us light years through history, science, psychology, art, folklore and his own travels, looking for an unpolluted dark and starry, starry night' IAIN FINLAYSON, *The Times*

THE END OF NIGHT

Searching for Natural Darkness in an Age of Artificial Light

PAUL BOGARD

FOURTH ESTATE • *London*

Fourth Estate
An imprint of HarperCollins*Publishers*
77–85 Fulham Palace Road
London W6 8JB
www.4thestate.co.uk

This Fourth Estate paperback edition published in 2014
First published in Great Britain by Fourth Estate 2013
First published in the US by Little, Brown 2013

A catalogue record for this book is
available from the British Library.

ISBN 978-0-00-742821-2

Printed and bound in Great Britain by
Clays Ltd, St Ives plc

To my mother and father.
And for all the life that depends on darkness.

To go in the dark with a light is to know the light.
To know the dark, go dark. Go without sight,
and find that the dark, too, blooms and sings,
and is traveled by dark feet and dark wings.

— Wendell Berry

Contents

THE END OF NIGHT

Introduction

To Know the Dark

Have you ever experienced Darkness, young man?

— Isaac Asimov (1941)

At least when it comes to light pollution, what happens in Vegas does not stay in Vegas. What happens here seeps across the surrounding desert so that national parks in Nevada, California, Utah, and Arizona, tasked with conserving their features "unimpaired for the enjoyment of future generations," report their horizons aglow, their dark skies tainted. It's to one of those parks, Great Basin, that I am headed—two hundred fifty miles north on Nevada's US Route 93, two lanes rising from I-15 toward Ely—to see for myself what's left of the dark.

The story is the same all over the country—dark places disappearing from the map. Computer images based on NASA photos show—from the 1950s to the 1970s to the 1990s—a steady spread of light across the land, and the projected view of 2025 imagines the entire country east of the Mississippi as one great rash of yellows and reds, the most intensely populated areas blisters of white; even west of the great river only scraps of black remain, each surrounded by a civilization gnawing at its ragged edges. Still, the eastern Nevada desert is some of the darkest geography left in the United States, and Great Basin National Park lies at its heart. So here I am, charging out of Las Vegas toward maybe the darkest spot in the nation.

It's the early evening, and all around the racing car the earth is changing, temperatures falling, animals and insects beginning to stretch and move, night-blooming plants feeling life surge again. All day the desert rocks have been gathering heat, expanding in sunlight, sending thermals skyward to soar hawks and bump descending planes. But at night the direction of energy flow reverses, the temperature drops thirty or forty degrees, and the desert rocks glow with warmth like a winter's woodstove. In the natural rhythm of day and night, whole mountains swell and fall like the chest of a sleeper.

To the east the mountain ranges still hold the rose color of the setting sun, while to the west already they are losing their definition, dissolving into silhouettes, the darkness sloping to the desert floor, long drapes hanging from mountainsides. We call this time "twilight," and officially, there are three stages—civil, nautical, and astronomical—that correspond to the gradual gathering of darkness and fading of the sun's light. In this twentieth-century classification, civil means the time when cars should use their headlights, nautical means dark enough that the stars needed for navigational purposes are visible, and astronomical means when the sky darkens nearly enough for the faintest stars. Unofficially, I love biologist Robin Wall Kimmerer's name for twilight: "that long blue moment."

We like to think that darkness "falls," as though it were like snow, but as the earth turns its back to the sun darkness actually rises from the east to wash and flood over land and sea. If you've ever stood at dusk and seen a gloaming on the eastern horizon, as though clouds were gathering, a thunderstorm brewing, that's what you're seeing—the earth's shadow as we rotate into it. What we call "night" is the time when we are caught in that shadow, a shadow that extends into space like the cone to earth's ice cream, a hundred times taller than it is wide, its vertex 860,000 miles above the earth.

Dawn comes as we rotate out of that shadow into the edges of direct sunlight.

Driving northeast away from what's left of this light, I look to the darkening sky and wonder what will be revealed. Venus, the Evening Star, emerges in the driver's-side window just over a silhouetted range, and then the first few actual stars, those of the Big Dipper, maybe the most well-known pattern of stars in the history of the world. One of these stars, Mizar, the second from the end of the Dipper's handle, is actually a double star, a visual binary, confirmed by telescope in 1650 but known to stargazers for millennia. In fact, the ability to see Mizar's faint twin, Alcor, with the naked eye has long been a traditional test of vision, one I admit I'm failing as the first bright town appears down the road.

The name of the town doesn't matter, for at least when it comes to light pollution this town is the same as ten thousand others: while its lights contribute only a little to the pollution blanketing the nation, all the different threads of the problem are here. The lights are all unshielded, for one thing, and so glare shoots this way and that, cast into the dark with little reason. Wood and chain link fences mark the boundaries between neighbors, but each neighbor's lights here, as all across America, are allowed to roam far beyond their boundaries—a perfect example of what dark sky advocates call "light trespass." The lights from these unshielded fixtures not only trespass onto the yards of neighbors and into the eyes of drivers passing through but straight into the sky, their energy wasted. The solitary gas station is lit beyond daylight, that light too floating from under the gas pump canopy to wipe stars from over the town. Drop-lens "cobrahead" streetlights are strung down every street, glaring into bedrooms and living rooms, the surrounding desert, and up toward the stars. Toward the edge of town come a smattering of "security lights," those ubiquitous white lamps hovering over backyards, barnyards, and driveways across the country, and then one final billboard lit from below, the upward-pointing light skipping from the ad into space without pause.

When the town ends, the darkness at its edge envelops the car, and my headlights cut the lit world down to just what lies ahead. The land on either side falls away, as though the highway is a bridge with thousand-foot drops to the left and to the right. The windshield soon resembles a Van Gogh night sky with its starry smattering of bugs. A jackrabbit sits eating at the side of the road, lifting its long ears absently as the car rushes past. Not long after, a coyote steps from the highway's other side, eyes aglow, a less fortunate jack hanging from its jaws. A barn owl lifts from a highway marker on the shoulder and flaps ahead for a few beats as though leading the way, then veers off into darkness and disappears.

In the Minneapolis suburb where I grew up lies a golf course with a road cutting through its center, white picket fence on either side. As a teenager I drove an old Volvo box that allowed me to turn off its headlights and sail a sloping, curving road lit only by parking lights, 35 miles an hour. The red wagon I own now is too smart and safe for that—the headlights remain on whether I want them or not—and I assume the same is true of my brand-new rental. But I'm wrong. The temptation is immediate and irresistible, and despite the fact that I'm not going 35 mph on this straight highway but nearly three times thirty-five, I rotate the dial.

In an instant the road disappears, my stomach drops, and I feel as though flung from the edge of the earth. The sensation is exhilarating fear, as my every fiber demands to know what I'm doing. I turn the headlights back on and feel my heart return to beating. The highway before and behind me holds no other cars, and no artificial lights shine in the black sea on the other side. I turn the lights off again and again—longer each time, long enough for my eyes to focus on what little of the highway my parking lights reveal, long enough to look ahead at the starry night flowing toward and over and past, and think of *Star Trek*'s starship *Enterprise* accelerating into space. Long enough to feel the car begin to float from the road's surface and fall into the sky.

The temptation is to leave off the lights, to drive in the dark for more than these few moments. But while I'm happy to know the thrill of boldly going 100 mph through the desert at night, to feel catapulted from earth into space, I am also happy to be alive, and so I slow to 20 mph. It's what seems now a trolling speed, and so I turn even the parking lights off and lean my head from the driver's window. The warm dry air flows over, the asphalt rolls underneath, and I realize I am headed directly toward a meeting at the horizon with the Milky Way as it bends from one end to the other. As though on its own, the car slows to a stop in the middle of Route 93 in the middle of the Great Basin desert. Any car or truck coming from either direction will show long before I'd need to move. Unless, of course, they are driving with their lights off, too, staring up at this altogether other highway.

"To know the dark, go dark," advises Wendell Berry. But seen from satellites at night, our planet's continents burn as though on fire.

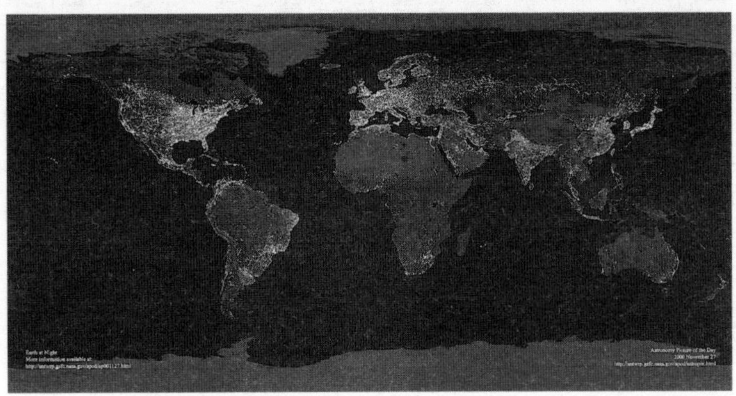

The earth at night, circa 2000. *(C. Mayhew & R. Simmon (NASA/GSFC), NOAA/ NGDC, DMSP Digital Archive)*

Across the globe the collected glow from streetlights, parking lots, gas stations, shopping centers, sports stadiums, office buildings, and individual houses clearly details borders between land and water, sometimes spreading even into the sea on squid fishing boats, their spotlights built to mimic noonday sun. It would be one thing if all this light were beneficial. But while some does good work—guiding our way, offering a sense of security, adding beauty to our nightscape—most is waste. The light we see in photos from space, from an airplane window, from our fourteenth-floor hotel room, is light allowed to shine into the sky, into our eyes, illuminating little of what it was meant to, and costing us dearly. In ways we have long understood, in others we are just beginning to understand, night's natural darkness has always been invaluable for our health and the health of the natural world, and every living creature suffers from its loss.

Our light-saturated age makes it difficult to imagine a time when night was actually dark, but not all that long ago it was. Until well into the twentieth century, what passed for outdoor lighting was simply one form or another of fire—torches, candles, or dim, stinking, unreliable lamps. And while these forms of lighting were an improvement on the earliest (skewering and burning oily fish or birds, gluing fireflies to your toes), how feeble this light was: A single 75-watt incandescent bulb burns one hundred times brighter than a candle. Historian E. Roger Ekirch reports that "premodern observers spoke sarcastically of candles that made 'darkness visible,'" and a French proverb advised, "By candle-light a goat is lady-like." Travelers considered moonlight to be the safest option for nighttime navigation, and lunar phases were watched far more closely than they are today. By the end of the seventeenth century, many European cities had some rudimentary form of public lighting, but not until the end of the nineteenth century did any system of electric lights—now so easily taken for granted—come into use. The darkness of our nights has been fading steadily ever since.

No continents burn brighter than North America and Europe.

Already, some two-thirds of Americans and Europeans no longer experience real night—that is, real darkness—and nearly all of us live in areas considered polluted by light. In the United States, Henry Beston's warning of "lights and ever more lights" from Cape Cod in 1928 may have seemed extreme for many of the 120 million Americans alive at the time, most of whom lived in rural areas without electricity, but fewer than ten years later he was well on his way to being proven right. With FDR's signing into existence the Rural Electrification Administration in 1935, the old geography of night in the United States was certain to change. By the mid-1950s, whether in the city, the suburb, or the country, most Americans lived with electric light. In the half-century since, as the American population has risen past 300 million, those lights have continued their steady spread unabated and, for the most part, unnoticed. Could we jump from the dark of the 1930s (or 1950s, or even 1970s) to that of tonight, few of us would fail to be impressed by the dramatic increase in artificial light. But that increase has been gradual enough that it would be easy to imagine our nights are still as dark, or nearly so, as they ever were.

With this in mind, and knowing, as he says, "the extent to which ever-growing light pollution has sullied the heavens," amateur astronomer John Bortle created in 2001 a scale on which he described various levels of dark skies, ranking them 9 to 1, brightest to darkest. He hoped his scale would "prove both enlightening and useful to observers," though he knew it might stun or even horrify some. While Bortle's distinctions can seem overly subtle, or inconsistent, they offer a language to help define what we mean when we talk about different shades of darkness, about what we have lost, what we still have, what we might regain.

Most of us are all too familiar with the brighter end of Bortle's scale—his Class 9: Inner-city Sky, or Class 7: Suburban/Urban transition, or Class 5: Suburban Sky—for these are the levels most of us call normal, what we call "dark." But Bortle's scale shows us what we are missing. Indeed, most Americans and Europeans, especially the youngest among us, have rarely or never experienced—and perhaps can't even imagine—a night dark enough to register 3 ("a rural sky" where only "some indication of light pollution is evident along the horizon") or 2 (a "truly dark site"). As for Bortle's Class 1, which he describes as a sky so dark that "the Milky Way casts obvious diffuse shadows on the ground," many question if such darkness still exists in the Lower 48. While rumors arrive from the deserts of eastern Oregon and southern Utah, the Nebraska prairie and the Texas-Mexico border, there's no denying that Bortle has described a level of darkness that for most of human history was common but for the modern Western world has become unreal.

From the moment I first encountered Bortle's scale, I wondered about the places I had visited and lived and loved, like the lake in northern Minnesota where as a child I first experienced real darkness and began to learn about night. I wondered as well if there were any Bortle Class 1 places left in my country. Another way to phrase that question is this: In the Lower 48 states, are there any places left with natural darkness? Or, yet another way, Is every place in my country now tarnished by light?

I decided to find out. I would travel from our brightest nights to our darkest, from the intensely lit cities where public lighting as we know it began to the sites where darkness ranking a 1 might still remain. Along the way, I would chronicle how night has changed, what that means, what we might do about it, and whether we should do anything at all. I wanted to understand especially how artificial light can be both undeniably wonderful, beautiful even, and still pose a long list of costs and concerns. I would start in cities such as Las Vegas—in NASA photographs the brightest pixel in the world—and in Paris, the City of Light. I would travel to Spain to explore "the dark night of the soul" and to Walden Pond to check in with Thoreau. I would meet with scientists, physicians, activists, and writers working to raise awareness of the value of darkness and the threats from light pollution: the epidemiologist who first connected artificial light at night with increased rates of cancer, and the retired astronomer who founded the world's first "dark sky" organization; the minister who preaches the necessity of the unknown, and the man whose work has saved countless nocturnal migrating songbirds in several major cities—it's through people like these that I would tell this story.

My first move was to contact Chad Moore, a founder of the National Park Service's Night Sky Team. For over a decade, Moore has been chronicling levels of darkness in the U.S. national parks, and I wanted to know what he thought I would find.

"Well," he explained, "as you slide down this ramp from nine to one into darkness it's not a smooth slide. It's... bumpy." Moore explained that with the Bortle scale, while the difference between 9 and 5, or between 5 and 2 would be obvious to anyone, the difference between 9 and 8 or between 2 and 1 can be difficult to discern. "There's so much fuzziness that it's prone to misinterpretation, and so if you're grumpy you'll give yourself a five, and if you're optimistic you'll give yourself a three... and it's really a four," he laughed.

That made sense, but are there still any Class 1 places left in the United States?

"There are rare places and rare moments where places in the U.S. compare with the rest of the world," he said. "I would like to think that I've seen that, that I've glimpsed Class One. But it takes some diligence. It's easier to get a plane ticket to Australia and drive out past Alice Springs.... It could take a while before you find that combination here in the United States."

Satellite photographs of the earth at night tell of two worlds—the illuminated civilization of developed (and developing) countries and the darkness of poor or uninhabited areas—and in some ways Moore is right; it would be easier to fly somewhere exotic and remote. But I wanted to know night closer to home. I wanted to know the darkness we experience in our daily life.

I decided to focus my journey on North America and western Europe. First, this is where the artificial lighting now sprawling over the world began and where it continues to evolve: It is Western thinking about darkness and light—and Western technology—that shapes the developed world's night. Second, few of us will ever fly to Australia and drive out past Alice Springs, but we all experience night where we live and work and love.

And most of us, if we wanted, could get ourselves out to real darkness closer to home, like the darkness of a rural highway in eastern Nevada.

"Our sun is one star in a disk-shaped swarm of several hundred billion stars," writes astronomer Chet Raymo. That disk-shaped swarm is our Milky Way Galaxy, and what arcs in three dimensions above this dark Nevada desert is the outer arm of that spiral, toward which we look from our inner-galaxy location. Raymo continues:

> I have often constructed a model of the Milky Way Galaxy on a classroom floor by pouring a box of salt into a pinwheel pattern. The demonstration is impressive, but the scale is wrong. If a grain of salt were to accurately represent a typical star, then the separate grains should be thousands of feet apart; a numerically and dimensionally precise model of the Galaxy would require 10,000 boxes of salt scattered in a flat circle larger than the cross-section of the Earth.

This means that every star in our night sky, every individual star any human has ever seen with his or her naked eye, is part of our galaxy and its "several hundred billion stars." Outside our galaxy exist innumerable other galaxies—one recent estimate put the number at 500 billion. At some quick point the size of the universe becomes overwhelming, its distances and numbers bending our brains as we try to comprehend the incomprehensible—that our night sky is but one tiny plot in a glowing garden too big to imagine.

But of course, for all of human history we have indeed imagined. Ancient civilizations from North America to Australia and Peru created constellations not only from groups of individual stars but even from the black shapes made by the gas and dust that lie between Earth and our view of the Milky Way's smokelike stream. And for ages we imagined it might well be smoke, or steam, or even milk—not until 1609 did Galileo's telescope confirm what he suspected, that the Milky Way's glow was the gathered light of countless stars.

In these countless stars, in their clusters and colors and constellations, in the "shooting" showers of blazing dust and ice, we have always found beauty. And in this beauty, the overwhelming size of the universe has seemed less ominous, Earth's own beauty more incredible. If indeed the numbers and distances of the night sky are so large that they become nearly meaningless, then let us find the meaning under our feet. There is no other place to go, the night sky makes this clear.

So let us go dark.

9

From a Starry Night to a Streetlight

It often seems to me that the night is much more alive and richly colored than the day.

— VINCENT VAN GOGH (1888)

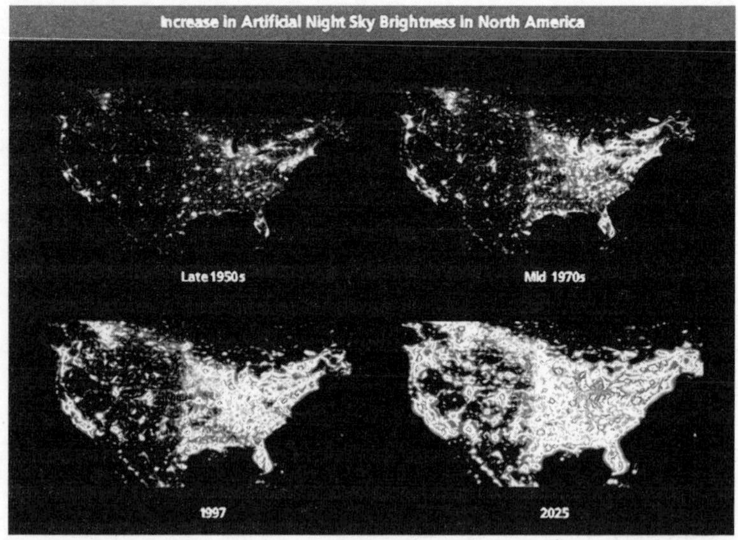

The growth of light pollution in the United States from the 1950s to the 1990s, and what light pollution might look like in 2025. *(P. Cinzano, F. Falchi [University of Padova], C. D. Elvidge [NOAA National Geophysical Data Center, Boulder]. Copyright Royal Astronomical Society. Reproduced from the* Monthly Notices of the Royal Astromical Society *by permission of Blackwell Science.)*

The brightest beam of light on Earth shoots from the apex of the Luxor casino's black pyramid in Las Vegas, thirty-nine brilliant blended xenon lamps, each six feet tall and three feet wide (the greatest number of lamps they could fit in the space), reflecting off mirrors and marking, like a push-pin on the night map of the known world, the brightest city on earth. New York, London, and Paris, Tokyo, Madrid, and a slew of cities in China—with their larger geographies and populations—may send more light into space overall than this single desert city in the American Southwest. But the "overall" qualifier is important, for it would be foolish to think there is any brighter real estate in the world than the Las Vegas Strip.

Standing on the corner of Las Vegas Boulevard and Bellagio Drive, I am immersed in artificial light, subsumed in the accumulated glow from the city's thousands of businesses and tens of thousands of homes, encased by peach-colored high-pressure sodium from the city's fifty thousand streetlights, most of which I'd seen from the plane just an hour ago. From the airport the Strip is only a short drive—the Luxor's beam on its south end meets you almost immediately—and in no time you are swallowed by light. Casinos rise bathed by floods, with ten million bulbs illuminating their glittering, flashing, changing signs. Digital screens and LED billboards call out from every corner, SEE OUR SHOWS! RENT OUR ROOMS! PLAY OUR SLOTS! Red lights, purple lights, green lights, blue—imported palm trees march past the illuminated iron footings of the Tour Eiffel of the Paris, Las Vegas casino, the tower drenched with gold-yellow light from base to tip-top, an exact replica of the real though half the size. A steady stream of headlights bob past, trailed by rafts of bright red tails. On a ruby-colored billboard truck, a blonde in a white bikini smiles. "HOT BABES, Direct to You." Most of the lights want to sell you something, and the Strip has the feel of one big outdoor mall, with canned music piped in and the natural desert pushed out. Some signs are brighter than others, some buildings more brightly lit, but everything is illuminated: The ground at my feet, the clothing on my body, the bare skin of my hands and arms and face, no surface remains uncoated—even the air itself seems full of light—and I walk through its presence as though pushing through an invisible scentless mist. In these first decades of the new century we live in a world that is brighter than ever before in history and growing brighter every year. If any city reflects that fact, it's Vegas.

Which is one reason I have come here to go stargazing. Rob Lambert, president of the Las Vegas Astronomical Society (there is such a thing, yes), has agreed to meet at the famous fountains in front of the Bellagio Hotel, saying, "I've got my telescopes in the back of my truck, so it will be no problem to bring them along." We may not have any luck—there can't be any better example than the Las Vegas Strip of Bortle Class 9, where "the entire sky is brightly lit, even at the zenith." But it's worth a try.

I wander over to the Bellagio, the tall curved casino set back from the reflecting pool housing the fountains, and when Lambert arrives we joke that we have chosen a popular spot, that we will be joined in our stargazing by hundreds of others—though they are here for comedians, magicians, musicians... and fountains—different stars than those we have come to see.

"People don't think about Las Vegas as a place to come look at the stars," Lambert tells me, "but we do quite a bit of outreach. Our slogan is, 'The greatest stars of Las Vegas can't be seen from the Strip.' Our club membership is only about a hundred, but when we have our star parties we have anywhere from seventy-five to five hundred fifty public."

Lambert takes out his laser pointer and cuts a thin green beam toward Orion—or, rather, the two bright stars from Orion we can see. "Okay, so that's Rigel on the bottom and Betelgeuse on the top left." He moves the laser lower to the left. "And there's Sirius, the brightest star in the sky." At first, I'm surprised we can see any stars tonight—this is my first visit to the Strip, and I had imagined the entire sky might be washed out by the lights. "Well, that's almost true," says Lambert. "When you consider that the stars we can see tonight are brighter than ninety-eight to ninety-nine percent of the stars our eyes could see, you start to realize what we're missing."

Behind us the water cannons begin going off, rumbling like distant thunder. The music changes to a kind of weird Italian carnival tune, coordinated with the booming cannons and joined by crashing cymbals. Someone nearby says, "I feel like breaking into song!" When I look to see who said this, I realize Lambert and I have turned so we face away from the fountain show, the only two in the crowd. "The winter Milky Way is actually over us," says Lambert, still looking at the sky, "but you can't see it..."

We agree to walk down the Strip to the Luxor, and as we start our trek south, Lambert tells me that he didn't get started with astronomy until after he turned fifty, that he'd heard some people at work talking about "star parties" and wondered what they were. Next thing he knew he was watching a friend's telescope at such a party, he says, and telling observers what they were seeing. "He had to go help someone with their scope and so he asked me to show people M13 through his telescope. So I said, sure, what's M13? He quickly told me M13 is a globular cluster in the constellation Hercules that is twenty-five thousand light-years away and made up of about seven hundred fifty thousand stars. And so for ninety minutes I told people everything I knew about M13, and absolutely had a ball."

We pass a man blasting solos on a cheap electric guitar, and further down the street the ghost of Keith Moon banging the hell out of a drum set, dozens of nude trading cards littering the sidewalk at our feet. On every block people are shouting into microphones, straining to be heard. Packs of partiers bump past, yelling their thoughts, half of them transfixed by cell phone screens, half staring, dazed at billboards pulsing light-emitting diodes (LEDs), and I'm reminded of how urban developers call signs like these "bug lights"—so bright they draw gawking crowds.

I ask Lambert about the appeal of looking at the night sky. "One of the things that I share with people is that, regardless of what your beliefs about creation are, it's still happening, there are stars being born, there are planets being born, stuff is still going on. For example, our 'challenge object' this month was Hubble's variable nebula, which changes all the time. You can look at it this year, look at it next year, and it's going to be different. And so you actually see things happening up there."

Not from downtown Las Vegas you don't, nor from any downtown in the developed world. While the lights of the night sky are far brighter than anything humans have ever created, all save one are so far away we see them as faint, if we see them at all. Instead, at night, we see the lights of our own making. While few cities have a space as intensely lit as the Las Vegas Strip, it's not just the Strip that makes Las Vegas bright. Here as in every city or suburb, it's the accumulated glow from an array of different sources that has utterly changed our experience of night.

During a recent Earth Hour—the worldwide movement in which cities are encouraged to turn off some of their lighting for an hour to draw attention to energy use—Lambert says he was driving on US Route 95 and was surprised at what he saw. "I was on 95 basically going from the north side of town to the south side, and this is where 95 is elevated above the valley. But when they cut the lights on the Strip there were so many streetlights that it didn't really affect the sky quality. You could tell that the Strip went dark because the hotels were no longer shining, but the quality of the sky didn't change."

In cities all over the world by far the greatest sources of light in our nights come from parking lots and streetlights (and, when in use, sports field lighting). While individually each streetlight might not seem so bright, it's together that they make their mark—in the

United States alone, some sixty million cobrahead streetlights blaze all night, every night, most still drop-lens high-pressure sodiums glaring their trademark pink-peach. We light our parking lots—think shopping centers, restaurants, hotels, stadiums, industrial areas, and the like—primarily with metal-halide lamps emitting intense white light. Add to these two sources a mix of auto dealerships, gas stations, convenience stores, driving ranges, sports practice fields, billboards, and residential neighborhoods and you have any given city's recipe for bright.

In general, bright lights lead to more bright lights, as with one corner gas station trying to outdo another. If you imagine a single light in an otherwise dark room, then turning on other lights around it, you see how the first light—bright in the context of the dark room—is now swamped, and in order to be noticed would have to become brighter. In Las Vegas, the ironic truth is that were the city's streetlights less numerous and less bright, the casino lights would actually appear more impressive.

Still, it's hard not to be impressed by the Luxor's beam, equal to the light of more than forty billion candles. In 1688, when the king of France decided to make a dramatic show of his power by illuminating Versailles, the Sun King shining in all his glory, all he could muster was twenty-four thousand candles. Granted, that is a lot of candles, and Versailles must have been beautiful—which is a word that at least for me is not so quickly applied to the Luxor's beam. But the intensity of this casino light is undeniable, and I can't help but stare. Though I'm staring too at what looks at first like sparkling confetti floating in the beam's white column.

"Bats and birds," Lambert says. "Feeding at their own buffet."

He's right. Dozens of bats and birds, drawn from desert roosts and caves, swooping and fluttering amid the casino's buffet of insects and moths attracted to the light. And how convenient, yes? Maybe not. In addition to the destruction wrought on the insects and moths, the Luxor's beam like a siren's song draws the bats and birds from their natural feeding habitat, causing them to expend so much energy flying to the casino that by the time they return from the journey they have nothing left to feed their young.

The sight reminds me of Ellen Meloy's essay "The Flora and Fauna of Las Vegas" and its concluding image: "Out from nowhere," Meloy explains as she stands outside the Mirage watching the casino's volcano erupt, "a single, frantic female mallard duck, her underside lit to molten gold by the tongues of flame, tries desperately to land in the volcano's moat.... Unable to land in this perilous jungle of people, lights, and fire, the duck veers down the block toward Caesars Palace. With a sudden *ffzzt* and a shower of sparks barely distinguishable from the ambient neon, the duck incinerates in the web of transmission line slicing through a seventy-foot gap in the Strip high-rises."

So bright and so recent—in evolutionary terms the Las Vegas we know today appeared in a sudden flash. The Luxor beam has only been on since 1993, several of the largest, brightest casinos have been built even since then, and the city's longest living residents were born before the first casino signs were illuminated in the mid-1940s. In less than a human lifetime what was almost an entirely dark place grew to the brightest place in the world, its population skipping from eight thousand in 1940 to sixty thousand in 1960 to more than two million today. "Welcome to Las Vegas," reads the famous sign, but only since 1959. Meloy's mallard, the bats and birds caught in the Luxor's beam? In terms of time to adapt, they've never had a chance.

As early as the mid-nineteenth century, some European cities were experimenting with electric light on their streets. As I walk past the Paris, Las Vegas, I think of an 1844 drawing showing a demonstration of arc lighting in the Place de la Concorde of Paris, France, cutting like a train's headlight through an otherwise jungle-dark night, catching in its glare the Place's fountains and a crowd in evening gowns and suits, some grasping umbrellas as protection against the light. Arc lighting was simply too bright for many uses, the first type of lighting that could truly be mentioned in the same sentence as the sun. (And it wasn't too bright simply because people had never seen anything like it. The moment I see a small arc light blazing away at the electricity museum in Christchurch, England, my immediate wish is for an arc welding mask—this was light that clearly could destroy your eyes.) As a result, it wasn't until 1870 that several European capitals installed arc lights on some main thoroughfares. While the intensity of these lights was so great they had to be placed on towers high above the streets, their arrival was met with fascination and pleasure by most (many cities in the United States barged ahead with their installation, for example). They were, it seemed, an answer to our prayers.

The idea had always been to banish darkness from night. As far back as the early eighteenth century, proposals had been made to illuminate the entire city of Paris using some kind of artificial light set high on a tower. The most famous of these was the Sun Tower proposed by Jules Bourdais for the 1889 Paris Exposition that would stand at the city's center near Pont Neuf and cover all of Paris with arc lights. Unfortunately for Bourdais (and fortunately for the rest of the world), his proposal was turned down in favor of one by a certain Gustave Eiffel. But even Eiffel's tower now has spotlights on the top, to the delight of some and the disgust of others.

Understanding how bright arc lights were, it makes sense how ready the world was for the incandescent light bulb. A report from the 1881 International Exposition of Electricity in Paris reads: "we normally imagine electric light to be a blindingly bright light, whose harshness hurts the eyes.... Here, however, we have a light source that has somehow been civilized." The change affected not only street lighting, of course. Arc lights had been entirely impractical for domestic houses, but as Jill Jonnes explains in *Empires of Light*, when incandescent bulbs arrived, wealthy, cultivated women in floor-length, rustling dresses delighted in showing their friends how if you just turned a knob on the wall, the room's clear incandescent bulbs began almost magically to glow, casting an even, clear light. Unlike candles, the electric light did not burn down or become smoky. Unlike gaslight, there was no slight odor, no eating up of a room's oxygen, no wick to trim or smoked-up glass globes to be cleaned.

It was in order to supply domestic customers like these that Thomas Edison in 1882 opened his first power station in lower Manhattan. By 1920 in America, electric service reached 35 percent of urban and suburban homes, and by World War II more than 90 percent of Americans had electric light. Still, it wasn't until FDR's insistence on the Rural Electrification Act of 1936 that electric light began to reach many areas of the rural United States, and not until well into the 1950s could one reasonably say most Americans enjoyed the benefits of electricity. Since that time, we have simply turned that knob on the wall farther and farther to the right, spreading electric light from city to city, town to town, up onto mountaintops and down into hollers, across the plains and into the desert, from coast to coast.

I sometimes try to imagine living in a city before electricity. How quiet pre-electric nights would have been without cars or trucks or taxis, without any internal combustion engines at all. No radios, televisions, or computers. No cell phones, no headphones, nor anything to plug those headphones into if you had them. How deserted the city with most of the population locked inside their homes, the night left to fears of crime, sickness, and immorality, and best avoided if one could. Finally, and most strangely—the biggest difference from that time to ours—not one single, solitary electric light.

How dark it would have been—imagine leaning out your door and, on the darkest nights, not being able to see more than a few feet in any direction. Historian Peter Baldwin describes as "downright perilous" the streets in early American cities, with few paved and then those only with cobblestones. On nights of clouds and no moon, he writes, "travel was obstructed along the sidewalks and street edges by an obstacle course of encroachments: cellar doors, stoops, stacks of cordwood, rubbish heaps, posts for awnings, and piles of construction material.... In 1830 a New York watchman running down a dark street toward the sound of a disturbance was killed when he collided blindly with a post." What lights did exist were intended only as beacons or guides rather than to illuminate the night. The New York street lanterns burning whale oil were in 1761 merely "yellow specks engulfed by darkness," and even more than a hundred years later its gas lamps were still "faint as a row of invalid glow-worms."

In *Brilliant*, Jane Brox tells of how American farm families, after they first got electric light, would turn on all the lamps in their house and drive out a ways just to watch it glow. Who can blame them? To go from the stink and dark and danger of kerosene to the clean well-lighted place brought by electricity—at the speed of light, no less. I would back away to admire the view as well. But soon it will be the rare person in the Western world who hasn't spent his or her entire life bathed in electric light, and no one will remember what night was like without it.

In the United States, our bright nights began with the first electric streetlight in Cleveland, April 29, 1879. But it was in New York City that the "possibilities of nighttime lighting first entered American consciousness," writes John A. Jackle in *City Lights: Illuminating the American Night*. "Once adopted there, its acceptance was assured almost everywhere." Thomas Edison returned to New York after a trip in 1891 to Europe proclaiming, "Paris impresses me favorably as the city of beautiful prospects, but not as a city of lights. New

York is far more impressive at night." Broadway was always the first, always led the way. It was the city's first street to be fully lit at night, first with whale oil lamps, then gas (1827), then finally electricity (1880). In a drawing from *City Lights* of Madison Square in 1881, arc lights on a tall pole shed light on an otherwise dark city scene of strolling couples, a horse-drawn carriage, telegraph poles and wires, and—in this famously windy part of the city—a man in the foreground who seems about to be blown over by the light as he crosses the street, cane in hand. By the 1890s, Broadway from Twenty-third Street to Thirty-fourth Street was so brightly illuminated by electric billboards that people began calling it The Great White Way.

These days, walking from lower Manhattan, it's not until I get to Thirty-first Street that I reach anything close to the bright white streetlights I had expected. Until then I'm in what feels, at least late on a summer Sunday, like a forgotten part of the city. With the theater district and advertising lights moved far up the street, the once bright Way is far more mild than Great, much less White than gray.

But once in Times Square, all that changes. Flashing digital signs, billboards, colored lights—from Forty-second to Forty-seventh is the brightest—and there is no night sky. I don't mean I can't see many stars, or even that I can't see *any* stars, I mean there appears to be no sky. Yes, above me, there is a blackish color—but with no points of light or any other indication of being alive. Instead, I feel as though I'm in a domed stadium. The light from the digital billboards simply drowns the white streetlights that lower on Broadway seemed so bright. I can honestly say it feels as bright as day. Maybe a cloudy day, but day nonetheless. Certainly, it no longer feels like night.

And by that I mean it no longer feels dark.

In fact, at least in terms of darkness, "real night" no longer exists in New York City, or in Las Vegas, or in hundreds of cities across the world. According to the World Atlas of the Artificial Night Sky Brightness, created in 2001 by Italians Pierantonio Cinzano and Fabio Falchi, two-thirds of the world's population—including 99 percent of people living in the continental United States and western Europe—no longer experience a truly dark night, a night untouched by artificial electric light. Satellite photographs of the earth at night show the dramatic spread of electric light over the globe—even without a map to show political boundaries, many cities, rivers, coastlines, and country borders are easily identifiable. But as impressive as these photographs are, they don't show the true extent of light pollution. Cinzano and Falchi took NASA data from the mid-1990s and, using computer calculations and imaging, showed that while in the photographs many areas outside cities appeared dark, they were actually flooded by pools of light spreading from the cities and towns around them. On the Atlas, levels of brightness are indicated by color, with white the brightest, and descending from there: red-orange-yellow-green-purple-gray-black. Like the NASA photographs before it, the World Atlas of the Artificial Night Sky Brightness has a certain beauty, but in truth it is a tale of pollution.

Light pollution is the reason Rob Lambert and I could see only a handful of stars from the Las Vegas Strip, and the reason I don't see any stars from Times Square. It's the reason why in the night skies under which the vast majority of us live, we can often count the stars we see on two hands (in the cities) or four (suburbs), rather than quickly losing count amid the more than twenty-five hundred stars otherwise visible on a clear night. It's the reason why even from the observatory deck of the Empire State Building we now see 1 percent of the stars those in 1700s-era Manhattan would have seen.

The International Dark-Sky Association (IDA) defines light pollution as "any adverse effect of artificial light, including sky glow, glare, light trespass, light clutter, decreased visibility at night,

A long exposure shows birds and bats hunting moths and insects amid the Luxor Casino's beam in Las Vegas. *(Tracy Byrnes)*

and energy waste." Sky glow—on display nightly over any city of any size—is that pink-orange glow alighting the clouds. It's tramping through a two-foot snowfall with the whole town bathed in push-pop orange. It's that dome of light on the horizon ahead though the sign says you've still got fifty miles to go. Glare is that bright light shining in your eyes that you raise your hand to block. Trespass is light allowed to cross from one property onto another. It's your neighbor's security light shining through your bedroom window. It's the lights on the brand-new science building that also illuminate the sororities across the street. It's all over every neighborhood in America, land of the free and the home of property rights. And clutter? A catchword for the confused lighting shining this way and that in any and every modern city.

The bad news? All mean wasted light, energy, and money. The good news? All are caused by poorly designed or installed light fixtures and our using more light than we need, and all could be significantly and—compared to other challenges we face—easily remedied.

When I think of how light pollution keeps us from knowing real darkness, real night, I think of Henry David Thoreau wondering in 1856, "Is it not a maimed and imperfect nature that I am conversant with?" He was writing about the woods around Walden Pond and how the "nobler" animals such as wolf and moose had been killed or scared away. "I hear that it is but an imperfect copy that I possess," he explained, "that my ancestors have torn out many of the first leaves and grandest passages, and mutilated it in many places. I should not like to think that some demigod had come before me and picked out some of the best of the stars." Some 150 years later, this is exactly what we have allowed our lights to do. "I wish to know an entire heaven and an entire earth," Thoreau concluded. Every time I read this I think, *Me, too.*

Bob Berman lives in a small town in upstate New York that has no streetlights. "I could never live in a place with streetlights," he tells me as we wind along a dark two-lane road, joined by a rising moon cast over the ruffled lakes and through bare spring trees, the songs of peepers audible over the sound of the car. We are on the way to the observatory he built by himself. Once described as the country's most popular astronomer, Berman has written a number of books, wrote the "Night Watchman" column for *Discover* magazine and the "Skyman" column for *Astronomy* magazine, and is known especially for his humorous writing style. Which is not the easiest thing to pull off when you're writing about astronomy, he says. "Science isn't inherently funny. What's funny about Pluto? What's funny about galaxies, and the cosmos, and the expanding universe? This is not social satire. When I was able to do a column on stupid questions, that was a gift from God."

"What was your favorite?"

"It's hard to top, 'If a solar eclipse is so dangerous, why are they having it.'"

But of course the "stupid questions" column had a serious point to make, that most Americans don't know much about the night sky.

I used to count myself among that number. I was always drawn to it, but I'd never known its names and numbers, its secret lives. In fact, here is what I did know: planets don't twinkle and therefore I could supposedly tell them from stars, and two prominent constellations—the Big Dipper (which technically is only part of a constellation) and Orion.

"That's not bad," Berman tells me. "The only thing most people know is the moon."

That I know more than I used to has a lot to do with Bob Berman, and especially his book *Secrets of the Night Sky*. Here's some of what I learned: One of the stars in Orion, Betelgeuse is "the largest single thing most of us will ever see." Sure, a galaxy is bigger, but a galaxy is a collection of stars rather than a single thing. Anyway, no galaxy is bright enough to shine through the light pollution that covers most of the developed world's skies. "Betelgeuse, on the other hand," writes Berman, "is brilliant enough to bulldoze its way through the milkiest urban conditions." Or how about this: Rigel, another brilliant star in Orion, "shines with the same light as fifty-eight thousand suns." Rigel is much farther away than the other stars making up the constellation, and, as Berman explains, if Rigel "were as close to us as the others, our nocturnal landscape would tingle with sharp, alien Rigel shadows, and the night sky would always be as bright as when a full moon is out. Most of the universe would disappear from view."

The moon tonight, a waning gibbous a few days past full, is bright enough that our view at the observatory won't be as great as it otherwise might be. When, during its twenty-nine-day cycle, the moon is big enough and therefore bright enough to wipe so many stars from view, most astronomers are not excited to see it. But Berman seems genuinely delighted to roll back the roof of his DIY observatory (which he built, he says, "crazy and wrong") and point his telescope at the moon. ("Did you make the telescope yourself?" I ask. "No, no, no," he says. "I wanted a good one.")

"Here, take a look at this," he says, and invites me to step up to the eyepiece.

I am not prepared for what I see: the gray-white moon in a sea of black, its surface in crisp relief, brighter than ever before. I am struck, too, by the scene's absolute silence. It is clearer, yes, brilliantly so, but this moon seems cold, antiseptic—alone in the unfathomable expanse of space. I learned a lot about the moon from Berman's writing ("it's more brilliant when it's higher, when it's nearer, and in winter when sunlight striking it is seven percent stronger"), and I appreciate this kind of information. But I think our relationship with the moon has more to it than simply astronomical facts. With my naked eye, on nights the moon climbs slowly, sometimes so dusted with rust and rose, brown, and gold tones that it nearly drips dirt colors and seems intimately braided with Earth, it feels close, part of this world, a friend. But through the telescope the moon seems—ironically—farther away.

"So now we'll go to Saturn," Berman says. Using both arms, he moves the large white telescope as though leading a dancing partner, turning it slightly to the east, then steps up the ladder and adjusts the view. "Now we're talking," he says. When I look through the scope, though, the bright tiny object is dancing around, and the image is blurry. Berman takes another look and makes some more adjustments, explaining as he does the key elements for viewing the night sky: transparency, darkness, and "seeing." Yes, that's what they call it. "You would think astronomers would come up with a more technical term," he chuckles, "but no, all around the world astronomers are saying, 'The seeing is a three point five tonight.'" Seeing reflects the effect of turbulence in the earth's atmosphere on the sharpness and steadiness of images—good seeing if the atmosphere is steady and calm, bad seeing if it is especially turbulent. A quick way to check seeing is how strongly stars twinkle: The more the twinkling, the worse the seeing. Berman tells me that bad seeing is one reason why no major observatories have been built east of the Mississippi for more than a century. The good "seeing" atop mountains in the desert has drawn astronomers west.

"Take a look at that," says Berman, climbing down. "Wait for the moments of good seeing—when it steadies up." Waiting for that—for good seeing—is exactly what an experienced observer will always do, he says. "I remember once when I was about twenty-four, and it was thirteen below zero, and my beard was frozen, I just stood there for three straight hours waiting for those moments when there would be steadiness and you could see ring within ring, detail that even photographs don't show. That's what observers have done for centuries."

As I wait for good seeing, I think about what Berman's just said. While humans have always watched the sky, modern astronomy has its origins in the lands we know as Egypt and Babylon, in the third and second millennia before Christ. People then were looking to the sky for signs and omens (though of course they were looking in other places, too; "the entrails of sheep were of special interest," writes historian Michael Hoskin). Eventually, the cosmology that developed—the classic Earth-centered Greek model of the universe—would dominate Greek, Islamic, and Latin thought for two thousand years. During the Middle Ages, astronomy in Europe was truly in the Dark, and not in the way modern astronomers would like—we have Islamic astronomers to thank for keeping the craft alive. That's the reason so many stars have Arabic names; one Islamic prince named Ulugh Beg, who lived from 1394 to 1449 in Samarkand in central Asia, catalogued over a thousand individual stars himself. And when, in 1609, Galileo Galilei (1564–1642) turned his handmade telescopes toward the sky, human observation of the cosmos changed forever.

When the "seeing" settles and Saturn comes into view, I can't help but say, "Oh my God!" To the naked eye Saturn is simply a bright starlike object—interesting, perhaps, but nothing more. But seen through a telescope it's a soft yellow marble with wide, striated rings—exactly as in photographs, but this time alive.

"Over the years more than a thousand people have looked through that telescope," Berman says. "And with Saturn, people always say one of two things. They either say, 'Oh my God!' or they say, 'That's not real!' "

That's not real—what a curious response. I've had other astronomers tell me the same thing, or say that people will question whether the astronomer hasn't placed an image of the planet into the telescope somehow. The fact that people are seeing something with their own eyes has incredible power—you can see photographs of Saturn a thousand times and be somewhat impressed, but see it for yourself and you don't soon forget.

The most beautiful starry night I have ever seen was more than twenty years ago, when I was backpacking through Europe as an eighteen-year-old high school graduate. I had traveled south from Spain into Morocco and from there south to the Atlas Mountains, at the edge of the Sahara desert, to a place where nomadic tribes came in from the desert to barter and trade, a place that when I look on a map I can no longer find. One night, in a youth hostel that was more like a stable, I woke and walked out into a snowstorm. But it wasn't the snow I was used to in Minnesota, or anywhere else I had been. Standing bare chest to cool night, wearing flip-flops and shorts, I let a storm of stars swirl around me. I remember no light pollution—heck, I remember no lights. But I remember the light around me—the sense of being lit by starlight—and that I could see the ground to which the stars seemed to be floating down. I saw the sky that night in three dimensions—the sky had depth, some stars seemingly close and some much farther away, the Milky Way so well defined it had what astronomers call "structure," that sense of its twisting depths. I remember stars from one horizon to the other, stars stranger in their numbers than the wooden cart full of severed goat heads I had seen that morning, or the poverty of the rag-clad children that afternoon, making a night sky so plush it still seems like a dream.

So much was right about that night. It was a time of my life when I was every day experiencing something new (food, people, places). I felt open to everything, as though I was made of clay, and the world was imprinting upon me its breathtaking beauty (and terrible reality). Standing nearly naked under that Moroccan sky, skin against the air, the dark, the stars, the night pressed its impression, and my lifelong connection was sealed.

When I tell Berman about Morocco he says, "A sad corollary to that story is when my wife's mom visited us once. And she had spent her life living in either Long Island or Florida, light-polluted places. We heard the car drive up, heard the trunk close, heard her wheel her luggage to the house, and when she came in she said to Marcy, 'What are all those white dots in the sky?' And of course Marcy said, 'Those are called stars, Mom.'"

"I've heard people say such things," I laugh, "but I can't believe they're true."

Berman leans back and calls, "Marcy, do you remember when your mom said what are those white dots in the sky?"

"Yep."

"Do you think she was kidding?"

"Nope."

Seeing stars is something Bob Berman has done all his life. And here in upstate New York, the sky still offers a wonderful view.

"We get down to about magnitude five point eight, five point nine, where you see a good twenty-five hundred stars," he says, referring to the scale astronomers use to describe the brightness of individual stars. "Theoretically, three thousand stars are visible to the naked eye at one time, but, in truth, since the overwhelming bulk of stars are fifth and sixth magnitude—the fainter you go, by far the more stars there are—and because extinction near the horizon is so great, the truth is faint stars stop at about ten degrees from the horizon and you lose a swath."

We adopted the idea of magnitude from the ancient Greeks, who called the brightest stars "first magnitude" and the dimmest "sixth magnitude." When modern astronomy put precise measurement to the Greek magnitude idea, a few of the brightest stars actually turned out to have minus numbers, such as Sirius (-1.5). But these values are all relative, reflecting only how we see these stars from Earth. The brightest star in the history of the universe could be fainter than faint if it's far enough from our view.

It's commonly accepted that the naked-eye limit is magnitude 6.5, though some observers report magnitude 7.0 or better. As Berman writes,

> There are few brilliant stars, many more medium ones, and a flood of faint stars. This hierarchy continues with a

> vengeance below the threshold of human vision. Recent advances have allowed telescopes to detect stars of magnitude 29—more than a billion times fainter than anything the unaided eye can perceive! This is very faint indeed: The light from such a star equals the glow of a single cigarette seen from 125,000 miles away.

But a light-polluted sky renders all this meaningless, of course, as the greatest wealth of stars lie in the larger values of magnitude, exactly the magnitudes erased by artificial light.

"My feeling," says Berman, "is that an observer needs to see four hundred fifty stars at a time to get that feeling of infinitude, and be swept away, and go, 'Oh, isn't that beautiful.' And I didn't make that number up arbitrarily—that's the number of stars that are available once you get dimmer than third magnitude. So in the city you see a dozen stars, a handful, and it's attractive to no one. In the city you say, 'Oh, there's Vega, who cares?' And if there's a hundred stars in the sky it still doesn't do it. There's a certain tipping point where people will look and there will be that planetarium view. And now you're touching that ancient core, whether it's collective memories or genetic memories or something else from way back before we were even human. So you gotta get *that*, and anything short of that doesn't do it."

Humans have long found in stars like these the familiar shapes that reflect our lives. For modern viewers these shapes sometimes make absolute sense, as with the scorpion of Scorpius, or the hunter Orion. The same is true of an asterism like the Summer Triangle, a shape made of bright stars from separate constellations: Vega from Lyra, the Lyre; Deneb from Cygnus, the Swan; and Altair from Aquila, the Eagle. But then there are the many constellations that puzzle us with their amorphous shapes and illogical identities, as though they are the eternal result of some ancient Greek joke. A good example is Auriga, with its bright star Capella, which lies just above Orion, readily apparent to any stargazer. How many of us would identify this shape as a charioteer? And Auriga is one of the easy constellations; try identifying Monoceros, the Unicorn, or Cetus, the Sea Monster, which both lie near Orion as well. With an image in mind (a good astronomy book or smartphone app helps), some of the ancient figures like Cassiopeia and Perseus may appear, but others (Ophiuchus, the Serpent Holder—which is hard to say, let alone see) are almost surely never to register.

Still, things could be worse, or at least more complicated. In 1627, the German astronomer Julius Schiller attempted to Christianize the sky by replacing the names of the constellations with the names of characters from the Bible. Thus the twelve constellations of the zodiac became the twelve disciples, and the constellations from the Northern Hemisphere were replaced by characters from the New Testament and those in the Southern Hemisphere with characters from the Old Testament. For better or worse, his idea never caught on. Not so lucky were the southern skies, where many of the constellations reflect the fascination of European explorers from the sixteenth, seventeenth, and eighteenth century for the practical new inventions of their time. Such inspirational constellations as the Air Pump, the Draftsman's Compass, and the Chemical Furnace live on today, as do the creatively named Telescopium and Microscopium. But not all is lost in the Southern Hemisphere, at least for children and those childlike at heart, all of whom can forever delight in pointing to the nautically inspired Puppis, the Poop Deck.

In order for us to see the stars in anything close to their possibilities—whether the awe-inspiring numbers Berman talks about, or in constellations both familiar and ridiculous—we need darkness. But how dark is dark? My hunch is that for most of us there are basically three levels. First, there is dark as in, "It's night, so it's dark." This is the standard notion of night's darkness, and it corresponds to around Bortle 8 down through Bortle 5. Next, at least for anyone lucky enough to find him- or herself in an area that would correspond to Bortle 4 or 3 or 2 (or, certainly 1), there is really dark, as in, "It's really dark out here." And finally, there is for some people a level of "darkness" that equates to "Vegas, baby!"

The reality, though, is far more complex. This is the message from Bortle's scale and the World Atlas of the Artificial Night Sky Brightness—we don't know what real darkness looks like, because we seldom ever see it.

One place to see real darkness in Manhattan is at the Museum of Modern Art, in Vincent van Gogh's *The Starry Night.* Unless Van Gogh's oil on canvas from 1889 is traveling as part of an exhibition, it hangs at home on its MoMA wall as fifty million people pass by every year. On a Saturday morning I stand near Van Gogh's scene of stars and moon and sleeping town, talking with its guardian for the day, Joseph, as he repeats, "No flash, no flash," "Two feet away," and "Too close, too close" again and again as people from around the world crowd near. "What's the appeal of this painting?" I ask. "It's beautiful," he says. "What more can you say than that?"

You could rightly leave it at that. But I love the story this painting tells, of a small dark town, a few yellow-orange gaslights in house windows, under a giant swirling and waving blue-green sky. This is a painting of our world from before night had been pushed back to the forest and the seas, from back when sleepy towns slept without streetlights. People are too quick, I think, to imagine the story of this painting—and especially this sky—is simply that of a crazy man, "a werewolf of energy," as Joachim Pissarro, curator at the MoMA exhibition Van Gogh and the Colors of the Night, would tell me. While Van Gogh certainly had his troubles, this painting looks as it does in part because it's of a time that no longer exists, a time when the night sky would have looked a lot more like this. Does Van Gogh use his imagination? Of course—he's said to have painted the scene in his asylum cell at St. Remy from studies he'd done outside and from memory—but this is an imagined sky inspired by a real sky of a kind few of the fifty million MoMA visitors have ever seen. It's an imagined sky inspired by the real sky over a town much darker than the towns we live in today. So a painting of a night imagined? Sure. But unreal?

In our age, yes. But Van Gogh lived in a time before electric light. In a letter from the summer of 1888, he described what he'd seen while walking a southern French beach:

> The deep blue sky was flecked with clouds of a blue deeper than the fundamental blue of intense cobalt, and others of a clearer blue, like the blue whiteness of the Milky Way. In the blue depth the stars were sparkling, greenish, yellow, white, pink, more brilliant, more sparkling gemlike than at home—even in Paris: opals you might call them, emeralds, lapis lazuli, rubies, sapphires.

It's remarkable to modern eyes, first of all, that Van Gogh would reference the stars over Paris—no one has seen a sky remotely close to this over Paris for at least fifty years. But stars of different colors? It's true. Even on a clear dark night the human eye struggles to notice these different colors because it works with two kinds of light receptors: rods and cones. The cones are the color sensors, but they don't respond to faint illumination. The rods are more finely attuned to dim light, but they don't discriminate colors. When we look at a starry sky, the sensitive but color-blind rods do most of the work, and so the stars appear mostly white. Add to this that we seldom stay outside long enough for our eyes to adapt to the dark, and then the fact that most of us live with a sky deafened by light pollution, and the idea that stars come in different colors seems wildly impossible, like something from Willy Wonka or Lewis Carroll (or

Vincent van Gogh). But gaze long enough, in a place dark enough that stars stand in clear three-dimensional beauty, and you will spot flashes of red, green, yellow, orange, and blue.

You may even feel as the Dutch painter did, that "looking at the stars always makes me dream."

But this morning at MoMA I am here to see two paintings, the second so little known that the museum doesn't even have it on display. It's through the kindness of Jennifer Schauer, who oversees the paintings in storage, that I get to see it. She marches me past *The Starry Night* to a room in which many paintings that the museum has no room to display are kept; 75 percent of the collection is here. Schauer looks at a label or two and then pulls out a fencelike wall on which the painting I've come to see hangs. And here it is, blazing away: Giacomo Balla's *Street Light* from 1909. For me, the fact MoMA has its view of a starry night on display every hour of every day, while this brilliantly colorful painting of an electric streetlight is hidden in backroom shadows, is deliciously ironic. This may be the only place in the city where the streetlight has been put away while the starry night continues to shine.

Here is a painting of the very thing that makes Van Gogh's vision of a starry night such an unrealistic one for most of us. In both paintings, the moon lives in the upper right corner, and for Van Gogh, the moon is a throbbing yellow presence pulsing with natural light. But for Balla, the moon has become a little biscuit wafer hanging on for dear life, overwhelmed by the electric streetlight. And that, in fact, was Balla's purpose. "Let's kill the Moonlight!" was the rallying cry from Balla's fellow Italian futurist, Filippo Marinetti. These futurists believed in noise and speed and light—human light, modern light, electric light. What use could we now have of something so yesterday as the moon?

"It's lighting itself up," Schauer says. On a canvas three times the size of *The Starry Night*, with a background of darkness painted sea blue-green and brown, the electric lamp radiates rose-mauve-green-yellow in upside-down Vs. The lamppost is a candy cane of those same colors, while concentric circles of the colorful Vs reverberate with resonant light. Here is an optimistic vision of what electricity would mean, not only a night brighter than what we'd known but one more beautiful as well. Indeed, were this what electric lighting had eventually come to be, Balla's reverence would be absolutely understandable even in our day. But of course, as my host says, "New York is never dark enough to see this."

And so here, fifty meters apart, hang two paintings that span a bridge of time when night began to change from something few of us have ever known into the night we know so well we don't even notice it anymore. Done in the southern French countryside at the end of the nineteenth century, Van Gogh's is a painting of old night. Done in the city at the start of the twentieth century, Balla's is a painting of night from now on. With time, electric lights like the one Balla portrayed would spread across western Europe and North America, perhaps inspiring the popularity of Van Gogh's painting as they did: As we lost our view of our own starry night, our view of his became more and more fantastic—this old night he had known and loved and experienced by gaslight.

8

Tales from Two Cities

The secrets are very simple. Blend light with the surroundings. Don't annoy the birds, the insects, the neighbors or the astronomers. If City Hall gave me money to do whatever I want, I'd teach people about the beauty of light.

— François Jousse (2010)

Gas street lighting first took flame on Pall Mall in London in 1807, with the light hailed as "beautifully white and brilliant." Within a decade more than forty thousand gas lamps lit over two hundred miles of London streets, a scene described by a visitor as "thousands of lamps, in long chains of fire." When, by 1825, the British capital was the most populous city in the world, no other place on earth was as extensively lit, or as bright.

Though "bright" depends on whom you ask. To the nineteenth-century eye—which until that time had never seen streets lit by more than candle lanterns or oil lamps—gaslight would have been unquestionably bright. But to our modern eye, gas lamps can seem questionably dim—you might wonder if they're even working. This isn't only perception—modern Londoners (as well as city dwellers all over the world, including 40 percent of Americans) live amid such a wash of electric light that their eyes never transition to scotopic, or night, vision—never move from relying on cone cells to rod. With gaslight, they did—the nineteenth-century eye saw gaslit nights with scotopic vision, and so what would seem to us incredibly dim seemed to a Londoner at the time the perfect artificial light, with "a brightness clear as summer's noon, but undazzling and soft as moonlight," one that created "a city of softness and mystery, with sudden pools of light fringed by blackness and silence." London ranks now as one of the brightest cities in the world—a white-hot splat on the World Atlas of the Artificial Night Sky Brightness. Nonetheless, I have come to the city to see if, even so, amid all that light, that "city of softness and mystery" remains.

I have a hunch it might, as London is still home to more than sixteen hundred gas lamps, most in the famous parts of town north of the Thames such as Westminster, the Temple, and St. James's Park. British Gas, which has direct responsibility for twelve hundred, employs a six-man gas lamp team made up of two gas engineers and four lamplighters, each of whom tends to four hundred lamps. Though they no longer need light each individual lamp, a task Robert Louis Stevenson described in 1881 as "speeding up the street and, at measured intervals, knocking another luminous hole into the dusk," the lamplighters make a circuit from lamp to lamp that usually runs about two weeks, cleaning the lamps, relighting pilot lights, winding the timers. Stevenson mourned the lamplighter's impending fate from the imminent arrival of electric light, writing, "The Greeks would have made a noble myth of such a one; how he distributed starlight, and, as soon as the need was over, re-collected it." Greek myth or no, the modern lamplighter's job is a popular one, with positions seldom changing hands.

On a crisp December evening, I join two members of the British Gas team, Gary and Iain, at St. Stephen's Tavern near Westminster Bridge, meeting amid the locals packed wall to wall, ties loosened and coats on their arms, before going out to see some of the best of the lights. About their affection for the lamps, both men are unabashed. "Once you get involved you fall in love with it," Gary says. And Iain, who moved to London from Glasgow, Scotland, tells me, "When I came down here I'd never seen a gas lamp before. I was totally taken by the history, and I found myself walking along the streets looking at electric lights thinking, *Bastards*, why is that not gas?"

London does seem to have an awkward relationship with its remaining lamps. While national heritage laws protect the lamps from being removed, apparently nothing in those laws protects them from being overwhelmed by electric light. On the edge of St. James's Park I saw what seemed to epitomize the situation. There stood a Victorian lamp fixture with perfectly good and glowing gaslight. Immediately to its right, less than two feet away, stood a taller, newer lamppost with a glaring electric light of no design but far more light. While certainly this back-to-back placement of gas and electric isn't uniform in London, rare is the street, park, or courtyard lit only by gaslight, and it's easy—if you're a fan of gaslight—to see this as an opportunity missed. As Iain says, "There's no doubt that electricity is a better way to do it, but you also can't deny the romance of gas lamps."

Both men consistently see evidence of the gas lamps' enduring public appeal. "You'd be surprised how many people walk by the lamps and don't bat an eye," says Gary. "But as soon as we go out and put up a ladder on them, everybody stops and starts taking photographs." Why do gas lamps appeal to us so? Part of it is simply that they are not as bright as electric lights—about as luminous as a 40-watt incandescent bulb. Part of it is that we like the Victorian fixtures; we like that style. Part of it is that gaslight's lower temperature offers the red-orange color of an open flame, which we're far more drawn to than bright white light. And finally, when you see a gas lamp on St. James Street in Covent Garden, or anywhere in London, you feel connected to the past—that nostalgia, that feeling of "so this is what it was like." Brightness, design, color, history—gaslight creates a beauty not better than electric light, but different.

One place to experience this truth best is around Westminster Abbey. The private courtyard behind the abbey known as Dean's

Yard is lit by gaslight, and, yes, it's much darker than you might expect from a city square. In fact, it takes a few minutes for our eyes to adjust after walking from the pub past Parliament and the abbey. But as our eyes grow used to the darkness, the light becomes perfectly bright. As Gary says as we look around, "You can see what you need to see. It's not daylight, but it's a lovely effect."

The effect, though, is subtle. If your objective is to light a football stadium, for example, you won't be happy with gaslight. When electricity first came to European streetlights, the public realized just how subtle the gaslight they'd grown used to was. You get the sense that some observers felt as though they'd been tricked. Said one Londoner, "Gas lighting had no effect whatsoever on the brightness of the street; it was not turned on at all for three evenings and nobody noticed the least difference." Said another, "as soon as we look away from the broad thoroughfare into one of the side streets, where a miserable, dim gaslight is flickering, the eyestrain begins. Here darkness reigns supreme, or rather, a weak, reddish glow, that is hardly enough to prevent collisions in the entrances of houses.... In a word, the most wretched light prevails." You can't blame nineteenth-century folks for feeling this way, and few of us today would willingly switch from electricity back to gas. But while we would never think to use gaslight to do the job of electric, we too often use electric lights to do the job of gas. You see this on Westminster Bridge, for example, where the lights seem far too bright, casting glare into the eyes of walkers, cyclists, and drivers. How much more beautiful would that bridge be if it were lit by flickering flames? And it's not that when we overshadow the gaslight we haven't a choice. If gaslight were deemed not bright enough for pedestrians, then waist-high, well-shielded electric lights could easily provide any walker the safety he or she needed while still allowing for the ambience created by the gaslights. Seeing London's gas lamps amid the city's already bright night raises the question of how we might genuinely appreciate all the benefits of electric light while at the same time avoiding what Stevenson called "that ugly blinding glare." In the face of such ugliness, he argued, "a man need not be very . . . epicurean if he prefer to see the face of beauty more becomingly displayed."

Stevenson wrote "A Plea for Gas Lamps" near the end of the nineteenth century, when arc lights were increasingly in use in Europe and the United States and the writing was clearly on the wall for gas lamps. His was not a tirade "against light" in general but rather a caution against what he saw as the uncontrolled, uncomfortable brilliance of the new electric lights. He wrote with admiration about how, with the coming of the "gas stars" in the streets, those streets were better places, and the lamplighters were good people, even if every once in a while, "an individual may have been knocked on the head by the ladder of the flying functionary." (I ask Gary about that. "Try not to," he says.) But now, with electric light, these lamplighters and their lights were to be replaced by "tame stars" that "are to come out . . . not one by one, but all in a body and at once," that is, with the flick of a switch. While wanting to "accept beauty where it comes," Stevenson cautioned against what he saw as "a new sort of urban star . . . horrible, unearthly, obnoxious to the human eye; a lamp for a nightmare!"

The technology we use to illuminate our nights has come a long way from those first arc lights, but I wonder if Stevenson's caution to us might be the same: While light at night is welcome, can there be such a thing as too much? In doing away with darkness, what beauty do we lose?

For hundreds of years this city was dark or nearly so, and I want to see if I can find more of that old London and the beauty it hides. I've chosen an old hotel in an old part of town where I can come and go by foot at any hour. For it's walking I have in mind, in the middle of the night, with Charles Dickens.

In "Night Walks" (1861), Dickens wrote, "Some years ago, a

temporary inability to sleep... caused me to walk about the streets all night." Dickens had walked "at half-past twelve" during the "damp, cloudy, and cold" London winter, figuring that with the sun not rising until half-past five, he had plenty of time to explore. I have come to London in winter as well, during the longest nights of the year, and when I wake at 2:20 a.m., I can't help but smile.

I dress warmly but go unarmed and without light—no flashlight, no headlamp, no torch. The hotel has some five hundred rooms, most booked, but none of my fellow guests join me as I trot down the stairs from the fifth floor. The hallways and stairs are as bright as in every hotel in the world and will remain that way through this middle of the night, this lost time after 1:30 a.m. until maybe 4:00 that feels more like yesterday and not yet tomorrow. In the lobby, I see no one but a custodian vacuuming near the front door's sliding glass. He doesn't see me until I'm almost to the door, then offers a look of *What? You're going out there?*

I'm first on the Strand—one of London's oldest and most well-known streets, on a cold December night, maybe 25 degrees—and then alone on Waterloo Bridge at quarter to three. West down the Thames, Big Ben and the Houses of Parliament stand dark against the gray charcoal London sky. Big Ben's round white face stands lit, as does the blue London Eye, the Ferris wheel on the river's south bank. Above, I count twenty-four stars. Behind, looking east, amid apparent smoke and steam, the unlit silhouette of St. Paul's Cathedral, the view a close copy of the famous photograph from the Blitz. That is, except for the skyscrapers going in behind, and the intense white lights off Blackfriars Bridge coming straight into my eyes.

Dickens describes the Thames as having "an awful look, the buildings on the banks were muffled in black shrouds, and the reflected light seemed to originate deep in the water, as if the specters of suicides were holding them to show where they went down." The river has claimed countless lives over the centuries, including those of eighteenth-century slaves jumping overboard to avoid their fate and six hundred passengers drowned in a paddle steamer sinking in 1878. I walk down the stony bank to the edge of the black water, the Thames feeling up close like a still wild presence at the heart of the frantic modern city. Tugs and barges and boats lie anchored, a yellow light in one illuminating a man coiling rope. Though the river still sees activity from police boats, fireboats, boats used by tourism companies, and—most significantly—barges used for civil engineering projects, times are changing quickly for the men who work on those barges that ply the Thames at night. In *Night Haunts*, Sukhdev Sandhu quotes a man who remembered that, when he was a child, "there were so many boats on the Thames that it was possible to skip from boat to boat all the way from one side of the river to the other without getting wet." These bargers, Sandhu writes, now "move through a river that appears to them to have been razed and colonized by outside forces, its soul abducted." Sandhu argues that while "Londoners take the Thames for granted... the bargers, especially after midnight, feel as though they have been unshackled from the city, its soot and heaviness, its noise and overbearing solidity. They breathe in the fumes of freedom, bathe in the tranquillity of the dark waters through which they gently move." It wasn't long ago, says Sandhu, that "the nocturnal river was swathed in darkness; now, even at its farthest reaches, car parks and grand shopping complexes are sprouting up, their light leaking out onto the Thames and denting its darkness."

Back west past Waterloo Bridge, I pass through an arcade of shops that was packed when I ran through this morning, my path a zigzag splash through puddles and past thick coats and couples and three-generation families. The embankments—this south side called "the Albert" and the north side called "the Victoria"—were built in the nineteenth century to control flooding by forcing the Thames to keep a set path rather than continuing its ancient seasonal weave. Now, this south bank is utterly deserted but absolutely lit. The only people I see are one security guard and one garbage collector. I make my way up and over Westminster Bridge, continuing along the South Bank toward Lambeth Bridge, looking across at Parliament. Until midnight, amber floodlights illuminate the Houses of Parliament, but in the middle of a winter's Sunday night, the old building stands dark from tip to tail. No lights in the windows, steam from only one among the many chimneys. With the clouds behind lit by streetlight glow, the building and towers stand in silhouette as though lit by moonlight centuries ago.

Dickens writes of crossing over Westminster Bridge and visiting the abbey, where he sensed "a wonderful procession of its dead among the dark arches and pillars." I feel the same way looking across the river at Parliament. By daylight, even floodlit, this is an old building in the present, but with its floodlight makeup removed, and placed in silhouette against the winter's sky, centuries fade and shadows come alive. Looking across the water, I imagine its ghosts coming down to the rooftop of the building in which they once walked. Whether you're in London or in the countryside or in your own bedroom, turning out the lights—especially the electric lights—can take you back in time.

From Westminster I walk to the corner of St. James's Park and around its curving boundary on Horse Guards Road, behind the Cabinet War Rooms and No. 10 Downing Street, crossing over The Mall and jogging up the steps past an enormous granite column topped by the bronze statue of a tremendously resolute Duke of York, and stop on Carlton House Terrace. If you want to see a street lit by gaslight, this is a good one—with no electric lamps in the way, the gas illuminates the street in soft golden flare. I continue on to Pall Mall and take a right down this famous old street, past rows of distinguished buildings, an open second-story window revealing a wall of ancient books—brown, crimson, black at the spine—and two third-story windows, drapes drawn, dim glow behind.

I think of Virginia Woolf and her essay "Street Haunting: A London Adventure," from 1927, in which she claims "the greatest pleasure of town life in winter" is "rambling the streets of London." Her story tells of using the excuse of needing to buy a pencil in order to get out and walk. "The hour should be the evening and the season winter," she says, because "the evening hour...gives us the irresponsibility which darkness and lamplight bestow." By irresponsibility, I like to think she means freedom. "How beautiful a London street is then," she writes, "with its islands of light, and its long groves of darkness."

I would like to have seen that London or, better yet, to see a modern London with "long groves of darkness" blooming with subtle "islands of light." Some eighty-five years after the publication of Woolf's essay, her equation has been reversed. Now, long groves of electric light give way only periodically to pockets of gaslit beauty or darkness. My visit here is the first time I experience a feeling that I'll have in several other cities and towns, especially in Europe, so rich with centuries of built history: how much more beautiful the nights could be here if more attention were paid to light and darkness. It's not that the London lighting doesn't have its moments—Parliament from across the Thames, for example—but in general the lighting relies on floodlights plastered against building walls, with the result a somewhat patchy appearance, especially compared to the subtle and more uniform lighting I will see in Paris. The opportunities for creating and enhancing the beauty of London at night are enormous—its gaslights and human history give it such an advantage over most cities in world—but for now these opportunities remain, for the most part, unrealized.

From Pall Mall to Trafalgar Square by 3:55, the "look right, look left" painted at intersection crossings, the black cabs slowing by, a row of sleeping red double-decker buses, Admiral Nelson immortalized with spotlights. Then, once again, the Strand. And lastly, Covent Garden.

This was the old market of London for hundreds of years, first outside the city walls, then at the city's edge. Dickens closes his essay by visiting "Covent-garden Market" which he finds on market morning "wonderful company. The great wagons of cabbages, with growers' men and boys lying asleep under them, and with sharp dogs from market-garden neighbourhoods looking after the whole, were as good as a party." The sense of a party in Covent Garden has a long history. An engraving from 1735 titled "Drunken Rakes and Watchmen in Covent Garden" features said rakes in tricorner hats, with swords drawn, their arms around ladies. A dog barks in the corner, lanterns lie smashed on cobblestones, the watchmen enter with their staffs a-swinging, giving one rake a serious kick in the behind, while a lady plugs her man's nose, and two link boys—who before gaslight "linked" travelers in the street from one lighted location to the next—stand holding their flames in the corner, clearly amused by ridiculous adults. What's interesting, aside from the crazy scene, is that in the background, modern-day Covent Garden is clear. Beyond the swooning face of a petticoated dame, you almost expect, if you look closely enough, to see the Apple Store logo in the shop window under the colonnade. The church with the clock, the clock tower, the passages, the cobblestones, they're all here. The description states, "He and his drunken companions raise a riot in Covent Garden," and funny enough, 275 years later, at just past 4:00 a.m., he and his drunken companions are still here, bellowing about Chelsea football as they lead each other, arms on shoulders, from the last pub to the next.

On the square itself, the gaslight still burns. But several shops are lit so brightly, either window displays or entire interiors, that electric light gushes into the square, flooding the night. To see how lovely Covent Garden used to be, stroll the side streets, Crown Court or Broad Court: gaslight and cobblestones, five-hundred-year-old buildings set close across from one another.

In Covent Garden Market now, night is still here, but morning is coming fast. It's time to walk back to the Strand, to my bed and sleep, and I have a strong feeling that when I return in a few hours the scene here will be changed. Back will be the shopping throngs, cups from Caffè Nero in one hand and shopping bags in the other, gone the ghosts of farmers, their cabbages, their dogs.

A few nights later, I am standing on Île St.-Louis, in the center of old Paris, watching the pale peach glow from nineteenth-century

A gas lamplighter in the Parisian darkness of the 1930s as captured by the photographer Brassaï. *(© The Brassaï Estate—RMN)*

lamps on a bridge over the Seine, the waxing crescent moon rising in a powder-blue lavender sky.

There are many bright cities, but only one City of Light, La Ville-Lumière. These days the city's nickname is often translated as "the city of lights," and with good reason, for the lighting of Paris is certainly part of its charm and identity. But if loads of electric light were all it took for a city to be called the City of Light, dozens of cities around the world would be well positioned to steal the title. We don't know the exact origin of the phrase, but we do know it refers to Paris's being the center of the eighteenth-century philosophical movement known as the Enlightenment. That is, the name City of Light has as much to do with new ways of thinking as it does with impressive artificial light.

It turns out that is still the case.

"Very little about this light is spontaneous," says David Downie, an American expatriate and author of the wonderful *Paris, Paris*, who has joined me for a walk through the old city. "It's all studied. Since 1900, they've consciously cultivated their image. Paris was really the first to pioneer this concept of a light identity. Of using light to create an atmosphere." Downie points to the lamps glowing on the short bridge from Île St.-Louis to Île de la Cité. "See the light fixture? It's a pre-1890s gas model with a little chimney, on a footbridge from the 1960s: That's what it's all about. They're playing not just with light but with shadow. The darker it gets, the better this bridge looks." While few would notice this bridge during the day, at night the lighting highlights the beautiful shadow-play on the bridge's underside. "There are a lot of little details that come out at night," Downie says. "They're very careful to make the light just strong enough so that you're not going to trip, but they've understood that they can't blind you. Here, they've created a nostalgic, old-fashioned feel with a warm blanket of light."

One feature of the lighting in old Paris is that there are few streetlights higher than fifteen meters, essentially no lights much above the first floor. The sidewalks and streets and balconies are lit, but above that the buildings fade toward darkness. "This was all studied; they want it to be this way," Downie explains. "The goal here is to create atmosphere, and the darker it gets, the more atmospheric it becomes."

As darkness collects between the buildings, along the Seine, on the rooftops, French doors, and balconies, rising around gold lamplight on the ancient narrow streets of these islands where the city was born, there is an intimacy, an openness—anyone can walk these islands, stand on these bridges, wander through this history, as though the city at night is a dinner party in a wonderful old house full of endlessly accessible rooms. The *fromagerie*, a little bell on the door and soft cheese white paper–wrapped, the *boucherie* windows full of twisted-neck fowl with feathered heads still attached, the Berthillon ice cream shop sending its small cones out into the night like messengers. Pipe organ notes sift through heavy, centuries-old wooden church doors, faux wicker-backed café chairs huddle around espresso-topped tables, a ribbon of moonlight ripples on the Seine's silver skin as it flows under bridges lit yellow-gold marching west toward the sea.

"This is the beauty of the night, a beauty 'rooted in atmosphere' that is not easily explained," explains Joachim Schlör in *Nights in the Big City*. "I start my nocturnal walk with pleasure, and my pulse beats slower in this pleasant darkness."

Walking and old Paris go together hand in glove, one reason so many Americans—used to cities enslaved to the automobile—revel in visiting the French capital. Recently the notion of the *flâneur* has gained popularity, one who appreciates, Schlör says, "the fine art of walking through a city slowly and attentively, one's appreciation bolstered by learning." But in Paris this walking happens not only during the day. In the *noctambule*, a word that in its English form, noctambulist, means sleepwalker but in French has a meaning closer to night owl, we find one who takes pleasure walking at night. The name was first applied to those Parisians who took advantage of the newly gaslit boulevards of the 1830s and 1840s, but for Downie, the quintessential *noctambule* is the eighteenth-century writer Rétif de la Bretonne. In terms of writing about walking at night, Bretonne paved the way, created the path. Downie often will follow Bretonne's old route as he walks around the edge of Île St.-Louis. "Bretonne lived right over there." Downie points. "He'd come out and walk the same way we're walking. He'd sit out at the end of the island and think great thoughts, and then go have his nighttime adventures."

During the years 1786 to 1793, Bretonne walked these streets of central Paris, and published his experiences in *Les Nuits de Paris.* That's only half the title, though. The full title—*Les Nuits de Paris, or, The Nocturnal Spectator*, by Nicolas-Edme Restif de la Bretonne—points to some of the pomp with which Bretonne carried himself. In a drawing on the opening pages of his book, he sports big-buckled shoes and stockings, a cape wrapped around himself, his hair falling to his shoulders, and upon his large wide-brimmed hat an owl (and this owl, with rabbitlike ears and wings spread, has a look of surprise, as though Bretonne has glued the bird's feet to his hat). Bretonne looks like a character—serious, thoughtful, and slightly ridiculous. And, in fact, that is how he reads. He came from Burgundy, which, at that time, was a totally dark place, and he couldn't get over the bright lights of the big city—in the 1780s, Paris suddenly had oil lamps, and more and more of them. "He was a mad walker," Downie explains. "He was completely bowled over by this idea that he could go out at night and walk around... and see."

This ability to go out at night and see, one we now so take for granted, had its origins in a decree by the French King Louis XIV in 1667 that lanterns be hung on Paris streets. As admirers proclaimed that "the night will be lit up as bright as day, in every street," the king commemorated his brilliant move by having coins minted featuring his profile on one side and, on the other, a statue of a robed figure holding a lantern and, in that lantern, a candle. And that—candles hanging over the streets of Paris—formed the first official system of public lighting in the world. By the end of the century, dozens of northern European cities had public lighting in their streets, some fueled with candles, others with oil. Paris alone lit more than five thousand candle lanterns, though only from October through March—the rest of the year relying on summer's lingering sunlight and the monthly advance of the moon.

Street lighting marked a dramatic change in human interaction with the night. Before this time, the coming of night's darkness signaled the end of working and socializing hours, the sign to come in from outside. As Wolfgang Schivelbusch explains in *Disenchanted Night*, "the medieval community prepared itself for dark like a ship's crew preparing to face a gathering storm. At sunset, people began a retreat indoors, locking and bolting everything behind them." To go out at night was to risk one's life, whether by a criminal's hand or a misplaced step—cables strung across the Seine caught the floating corpses of those who had fallen off the quais or bridges and drowned in the dark. The new public lighting facilitated and acknowledged a changing culture. Coffeehouses were spreading through northern Europe and cafés were staying open later and later, marked by a lantern hanging over the door. Along with stronger state security, these increased opportunities for socialization and commerce joined with the new lights to open the darkness to more and more people. Eating, drinking, working—this opening of nighttime hours radically altered life for northern European city dwellers. By 1800, for example, mealtimes had shifted back by as many as seven hours from those of the Middle Ages. "Nocturnalization," historian Craig Koslofsky calls these changes, the "ongoing expansion of the legitimate social and symbolic uses of the night," was an expansion for which street lighting served as infrastructure.

By midcentury the candle lanterns in Paris had been replaced by a new type of oil lantern, the reflector, or *réverbère*, which used multiple wicks and two reflectors to produce dramatically increased amounts of light. In fact, *réverbères* were enthusiastically hailed as artificial suns that "turned night into day." A report prepared for the Paris police chief in 1770 suggested, "The amount of light they cast makes it difficult to imagine that anything brighter could exist." But for eighteenth-century Parisians it didn't take long for the shine to wear off. "These lights cast nothing but darkness made visible," wrote one Frenchman. "From a distance they hurt the eyes, from close up they give hardly any light, and standing directly underneath one, one might as well be in the dark." Indeed, a century after the Sun King's decree, an Englishman visiting Paris declared, "This town is large, stinking, and ill lighted."

For any *noctambule*, Paris held plenty of challenges. The narrow streets had no sidewalks, and death by stagecoach was not an infrequent event. "There are nights when all the disadvantages of a crowded quarter are apparent at the same time," Bretonne wrote. "As I was coming off rue du Foarre, a large marrow bone fell at my feet. Its sharp force and the force with which it was hurled would have made a lethal weapon of it, had it struck me." As his walk continued, he faced a "sheet of soapy water" thrown from a window, then a bucket of ashes. Still, things could have been worse. The city's dirt and pebble streets were lined with sewage and waste, the air filled with a rank stench that we could only imagine by standing in a town dump. Writes historian Roger Ekirch, "The Duchess of Orleans expressed amazement in 1720 that Paris did not have 'entire rivers of piss' from the men who urinated in streets already littered with dung from horses and livestock. Ditches, a foot or more deep, grew clogged with ashes, oyster shells, and animal carcasses," and "most notorious were the showers of urine and excrement that bombarded streets at night from open windows and doors." William Hogarth's painting of London in *Night*, from *The Four Times of the Day* (1736), might just as well have been a city street scene from Paris: A woman pours a bucket of human waste out an open window onto the back of an unfortunate man who staggers along with his wife. He holds a stick and she a lantern and sword. Oh, and a bonfire burns in the middle of the narrow street behind them.

Amid this dimly lit craziness it might be difficult to believe that street lighting could be a source of bitterness and anger. But in the years before the French Revolution, street lighting was often a thorn in the public's side. From its start, public street lighting had been significantly motivated by the state's desire to gain control over the streets at night, and for many Parisians the oil lamp simply stood for tyranny. When lanterns were at first hung low, they made for easy targets, destroyed with walking sticks. But when the lanterns were then hung out of reach, a new technique emerged, that of cutting the lantern's ropes and letting the lantern smash into the street. At times, like the modern-day smashing of Halloween pumpkins, smashing lanterns was simply a form of entertainment. As Schivelbusch writes, "Whatever the details and methods, smashing lanterns was obviously an extremely enjoyable activity."

While the candle lanterns and *réverbères* are long gone, and electric lighting makes Paris today as bright overall as any city its size, the echoes of such history remain. Though some complain that old Paris has become a museum or even that it's dead, I think it anything but, and I think that especially at night. What's kept alive is the opportunity to add your story to those countless stories before, even to add to your own story if you have been here in the past. Because so much of the old city has been preserved, you can come back to Paris and the night you walked years ago will still be here.

The year after high school, while backpacking nine months through Europe, I remember especially a week in Paris, in the winter, alone. I had lucked out and discovered a small hotel on the Île de la Cité, the Henry IV on the Place Dauphine. I would set out each night and walk for hours through old Paris, the gray-black Seine there to guide the way, the long, gray ministry buildings with rooftops black and windows dark, the French tricolor spotlit in front. I would stand on the Pont Neuf and wonder where this life would lead.

As Downie and I continue our stroll, I tell him that the night I reached the city this year, I took the métro from the Gare du Nord to the Champs-Élysées, near the Arc de Triomphe. A wet snowfall weighed on tree branches and café awnings, crystals sparkling. The snow snarled rush-hour traffic and slowed walkers with its slushy challenge, casting a hush over the sounds of wet tires and boots. The leafless smooth-barked plane trees along the avenue were filled with small white lights, bright with a tinge of sky blue, each with two or three long bulbs like fluorescent ceiling lamps dark except for a periodic slide of light down their length, the movement like one of melting snow sliding down a rock face or roof. At the end of the wide Champs-Élysées boulevard, I wandered past the bright blue-white Ferris wheel (La Grande Roue) set up for the season in the enormous Place de la Concorde, the famous city square where King Louis XVI lost his head to the guillotine and where the spotlit Obelisk—a 3,300-year-old Egyptian column—stands seventy-five feet tall. From the Place I skirted around the locked and deserted Tuileries gardens, which, during the day, fills with couples and families and solo strollers, and found myself among the stone buildings of the Louvre Palace, where black lampposts ring the courtyard with bright light. Then, along the Seine to the Île de la Cité, past the large Christmas tree filled with navy-blue lights in front of Notre-Dame, around the cathedral onto Île St.-Louis, and through the amber-lit Marais neighborhood to my hotel. All told, a walk of nearly two hours, but in that time I saw much of the old city. No museums or galleries or music or events, not even a glass of *vin rouge* or a quick stop at a *créperie*. But the City of Light on a dark winter's eve and nearly for free, the priceless sensation of having returned to something once mine.

"Everything belongs to me in the night," wrote Bretonne. In Paris more than two hundred years later, that truth remains: Everything is accessible, at least to your eyes—monuments, famous buildings, ancient streets. Little is closed off as you walk this city at night. Even—as the lights come on in apartments you pass—other people's lives.

Downie nods. "My wife says it reminds her of an Advent calendar, the way the windows suddenly come to life." We've reached the Place des Vosges, built in the early 1600s, the oldest planned square in the city, lined with grand two-story apartments. "That's a seventeenth-century painted ceiling," he says, pointing. "This city is full of unbelievable interiors, and you only see them at night."

In a neighboring apartment are long maroon drapes pulled back from French doors, the lifted slant top of a grand piano, and in the corner on the wall a stag's head. "Now, speaking of expensive," Downie says, "this is a double pavilion owned by one man, one very rich family. They've owned it for a hundred seventy years. And if you look, see that tapestry? It's a sixteenth-century tapestry. If other lights were on, you would see amazing things because he's one of the most successful auctioneers in the country."

These are rooms into which, during the day, I would never be allowed. But at night, walking Paris, invited into room after room, life after life, I feel welcomed to enjoy the beauty this city offers. And I want to know more.

François Jousse emerges into the Parisian evening as though from out of the shadows, ambling toward me from behind the enormous Christmas tree in front of Notre-Dame. With his bushy beard, red plaid coat, and camel-colored hiking boots, he looks like a lumberjack. It turns out that those boots are key—Jousse likes to walk Paris, day and night, and that's what we have agreed to do: a tour of central Paris so he can tell me about his work. He is immediately jovial, friendly and cheerful, clearly delighted to be talking about lighting the city he loves, albeit slowly in English with a heavy French accent. He begins many of his sentences with "*Alors*...," meaning "So . . . ," before explaining something new. There is much to explain, because there has been so much thought put into the lighting of Paris. And the man who has done much of that thinking, the man who has done so much to create the atmosphere of Paris at night, is François Jousse.

We start at Notre-Dame, where in 2002 Jousse oversaw the completion of a ten-year, multi-million-dollar upgrade to the cathedral's exterior lighting. For several decades after World War II the cathedral was simply spotlit, and then only its façade. Before the war, it had spent centuries in darkness—a Brassaï photograph from the early 1930s, shot from Île St.-Louis, shows the cathedral from behind, lit only by surrounding streetlights, a dark hulking shape as though carved from coal. Not until recently—not until Jousse—did the city take seriously a relighting of one of its most enduring landmarks. "For the lighting of the cathedral we made a competition, a jury with clergy, cultural minister, city of Paris—many people," he says with a slight grin, "and it was very, very difficult." Jousse tells me one idea was to have the cathedral's famous stained-glass rose window lit from within, a proposal of which the clergy disapproved. "They said," Jousee laughs, "we were the devil."

For Jousse, the project of lighting the famous cathedral didn't stop with just the church. When he says "the cathedral," he explains, he means not only its face but everything around the building, the lights of the bridge adjacent to it, the plaza in front of the cathedral. "The concept was to put the cathedral in the center of the island. And to tell a story." For example, Jousse points out how the lighting grows gradually brighter as it reaches the cathedral's top, intentionally drawing the viewer's gaze skyward—toward heaven. And though pleased with the project, Jousse says he didn't get everything he wanted. "I have made also a design for this garden," he explains as we pass the dark courtyard behind the cathedral, "but no money." He then offers a "what-can-you-do" laugh, lowers his gaze, and we're off again, walking to the next stop, stretches of silence but for the crunch and splash of our boots in the snow-crust and melt-slop of the Paris sidewalks.

Speaking of money, the city now spends some 150,000 euros each night for the electricity, maintenance, and renovation of its lighting, a quantifiable reflection of its commitment. But this wasn't always the case. When Jousse took his position in 1981, Paris at night looked little like it does now. As with Notre-Dame, the city's famous monuments and buildings were mostly spotlit, and many others were not lit at all. Over the course of thirty years, Jousse and his associates relit Paris almost entirely—more than three hundred buildings, thirty-six bridges, the streets and boulevards—all with the goal of integrating them into the city, being as economical as possible, and creating beauty. Before his retirement in 2011 as chief engineer for doctrine, expertise, and technical control, Jousse was the man in charge. His car even held a special permit that allowed him to park wherever he wanted in order to better troubleshoot, direct, or otherwise consider how Paris would be lit.

Most visitors to Paris probably notice the beauty of the lighting, but they probably don't notice how carefully that beauty is created—where and how the floodlights are placed, the challenges the lighting designers faced, the amount of energy used. That's just fine for Jousse. In fact, he delights in showing me how he hid many of the projectors so that the lights become part of the building, and the building part of the city. He doesn't want to draw attention to the lighting, nor does he want the lighted building to stand out from the neighborhood. On the sidewalk across the Seine from Notre-Dame, at the end of a long row of green metal stalls—those of the famous *bouquinistes*, the booksellers whose presence here began in the 1600s—Jousse shows me how the first two stalls actually house no books, hiding two spotlights instead. Anyone walking past the bookstalls would never guess the light on the cathedral came from within them.

"Whose idea was this?" I ask.

"This was mine." He laughs.

Jousse sees himself as a historian of technique, and a storyteller using light as his language. As we walk past the Hôtel de Ville he says, "Now I show to you my last design in Paris." He leads me toward the Tour St.-Jacques, the 170-foot Gothic tower that is all that remains of the wonderfully named sixteenth-century Church of St.-Jacques-de-la-Boucherie (St. James of the Butchery). Jousse used the story of Blaise Pascal's experiments with atmospheric pressure as inspiration to develop this lighting design. "I want to make homage to Pascal. The light falls from the top, and when it reaches the ground it makes a splash." And indeed, the light starts brighter at the top, fades as it falls, then brightens and spreads at the tower's base. This blend of artistic thinking with technical solutions essentially describes Jousse's work in Paris—to think about the philosophy behind the light, and then to make it happen. "I want that the building says something with the light," he explains. "But the speaking can be different. Maybe it's an architectural speech, maybe it's a historical speech, maybe it's humorous. Sometimes the speech can be spiritual. Sometimes people say to me, But nobody will understand what the building says. And I say, It's not a problem, the building says something and it's beautiful because the building says something."

At St.-Eustache I see what he means. From a block away the cathedral seems to rise from darkness, its bottom half left unlighted, its top half glowing subtle amber-gold. Jousse smiles. "For the church I want to be sure the light says something. I give the speaking to one designer, and the technique to another. And the first one must say, 'I see the church like that during the night because na, na, na, na,'" he laughs. "It was the first realization with this way of thinking, maybe in the world. And his speech was something like, the church is like a battery of God-energy. During the day the church takes in the energy of God, at night the energy of God comes from inside to go outside."

As we walk closer to the church the bottom half emerges from the shadows, its stone arches lit by ambient rather than direct light. "When you're far away you ask why isn't it lit, but when you're up close you don't ask anymore," he says, clearly satisfied. "There's comfort, and there's ambience. Everything doesn't necessarily have to be lit. On the contrary, it's when you leave things in shadows that you see the light better."

I wonder if the same could be said about light and quiet.

The sounds of city traffic fall away as we walk into the Louvre courtyard called the Cour Carrée, a small square with a circular fountain in the middle, and on the three stories of sandstone and windows the golden light of some 110,000 small (4.5-watt) lamps ("the same number as all the other lamps in Paris," Jousse explains). "It's very beautiful," he says, this time more serious. "C'est magique." The effect created is that rather than the lights shining on the building, the building seems to be emitting the light. "The picture is fantastic. The maintenance is also fantastic." He laughs. The energy for this one courtyard alone costs one million euros per year.

We leave and cross the busy street to a bridge, Le Pont des Arts. "Et voilà," Jousse says. "Another magic area of Paris." Yes, this one a romantic pedestrian overpass made of iron and wood. Jousse says the challenge here was that on this slim bridge there were no good places to put light projectors. "It's a very poetic place," he says, "and if people have projectors in their eyes it's not good. But the city says to me, 'All bridges must be illuminated.' So, I say okay." He chuckles. Jousse solved the problem by placing his projectors under the bridge facing the river, and illuminated the bridge from the light reflecting off the moving water, thus creating a shimmering, beautiful effect.

What does it mean, I ask, to include values of beauty and poetry and love when you're working with light? "It's hard for me to answer," he says, "I'm an engineer, not a poet. But as far as love goes, I would say that's true. Oui, c'est vrai. I'm in love with Paris." He laughs. "If you work on lighting without having any love for what you're lighting...," he trails off, as though there's nothing more to be said. Then: "The love of Paris comes first, the lighting of Paris is secondary."

For our last stop we take the métro up to Montmartre and look down onto the city, the softly lit white curves of the Sacré-Cœur church behind us (another of his lighting designs? *Oui*). The Eiffel Tower stands over the dark city, lit from within by three hundred fifty sodium vapor lights designed to mimic the amber glow of the gas lamps that once lined the interior of the structure. Only three decades ago, just one side of the tower was lit, all the lighting from spotlights stationed by the Trocadéro Palace. Jousse tells me that the energy consumption was huge, and because of the tower's brown paint you couldn't see any details. Then came the idea to light the tower from within. Since then, except for each top of the hour, when twenty thousand white lights sparkle the tower for ten minutes, or on rare occasions (briefly all in red for a visit from the Chinese premier, all in blue to honor the European Union), the lighting hasn't changed for twenty-five years. "And for us it's very conservative, it's classical. It's beautiful like a jewel, but it doesn't change. But it could be worse; it could be a wedding cake. So, sometimes classical is good."

When I share my appreciation for the role lighting plays in the story Paris tells, he says, "If you feel that way, then I am very happy." With this, Jousse bids me farewell.

I turn and look out over Paris. From Montmartre, you see the pollution from the suburbs at the edges of the city, their butterscotch orange lights running unleashed into the sky. But the old Paris looks dark, the view a direct result of the rules that light fixtures be directed downward and the lights themselves not be placed any higher than they are. The effect is that of an old city in preindustrial darkness, though under that canopy you know there lives and breathes a city of light.

When I turn back toward Sacré-Cœur, François Jousse is rounding a corner of the church, his head lowered, his boots returning him to the shadows.

7
Light That Blinds, Fear That Enlightens

After thousands of years we're still strangers to darkness, fearful aliens in an enemy camp with our arms crossed over our chests.

— Annie Dillard (1974)

Rolling hills, gnarly old trees, a creek running through—when I return at Christmas to the suburban Minneapolis neighborhood where I grew up I wait until just before midnight, then head with my dog Luna two blocks south, slip through a tear in the chain link fence, and take a golf course walk. On account of liability fears we're not supposed to be here. But we are, and it's a pleasure, walking in what passes for dark. The city-lit sky and snow-swamped land combine—darker than day, but lighter than night ought to be. The leafless limbs of oaks and maples and the nests of birds and squirrels high in the branches, against the glowing winter sky, are like x-ray images of various animals, of vascular systems and hearts. Some years, solitary owls perch in silhouetted trees, watching me until I notice, then swooping away. Other years, deer crossing a fairway in the distance, or the circular squeal-yipping-bark of coyotes by the railroad tracks. And once, looking back, the weightless drifting prance of a fox crossing the snowy sloping hillside we'd just tread.

To the east the city rises in golds trimmed in royal blues and sparkling reds, silvers, and whites, steam twisting street-level to sky. Sky glow colors the entire eastern horizon hazy orange—and with the south, west, and northern horizons all gray-white, any low-hanging star has been wiped away. Only overhead are maybe four dozen, no more—Orion; the Pleiades; Sirius, the Dog Star. It seems like night here but it's not, at least not as it would be without all this light.

Slipping back through the fence, walking home, we are bathed by corner streetlights and the 100-watt bulbs in "brass and glass" front-door fixtures. The combination of house lights and streetlights and city-supplied sky glow illuminates the four blocks to the street's end, each house defined. It's a scene repeated in every direction and, with rare exception, over the suburb as a whole. It's the kind of suburb in which tens of millions of Americans have grown up learning what "dark" is, the kind of suburb in which one hundred million Americans live. You would never see the Milky Way here, or meteors, or anything close to Van Gogh's wild night, and on Bortle's scale, on its darkest nights, this suburb would be lucky to rank a 7. And still, a few years ago, the people on this street asked for more light.

In the forty years my parents have lived here, there has never been any trouble with crime. That is, the type of crime we fear—the stranger snooping outside the window, sneaking in the back door, doing us harm. Even so, the neighborhood petitioned the city government, and soon five straight metallic poles topped by yellow carriage fixtures had been stitched into the street at fifty-yard intervals. From one night to the next, gone was what had been left of the street my mother had chosen because it reminded her of the dark country roads in Ohio where she'd grown up in the 1950s. "I was against it," she says of more streetlights, "but I was outvoted."

Why? I ask.

"Oh," my father says. "Safety and security."

Sooner or later, when talking about artificial lights and darkness, you come to questions of safety and security. Usually, it's sooner. In fact, the first question at any presentation about light pollution is bound to be something like, "Yes, so it's great to see the night sky and everything, but we need lights for safety." This isn't actually a question, I realize, and usually the speaker isn't really asking but rather stating what we have all been taught is fact. But often that statement has a subtext, too, something like what I found on a Colorado website: "less street lighting means more rapes, more assaults, more robberies, and more murders. It is wonderful to be able to see the details of the Crab Nebula from your back yard. It is also wonderful to be able to walk down the street without being attacked by a violent predator."

You don't have to look far to find the idea that darkness and danger go together, as do security and light. In Oakland, a city with thirty-seven thousand streetlights, an assistant police chief claims increased lighting levels could help reduce crime because "most of these crooks, when they commit a crime, want to do it in darkness." In Boston, with sixty-seven thousand streetlights of its own, a Northeastern University criminology professor argues that lights act as "natural surveillance" and can reduce crime by 20 percent. In Los Angeles, home to more than two hundred forty thousand streetlights, the city attributes a 17 percent drop in violent gang-related crimes in the areas surrounding parks to those parks' having received new lights. And here in Minneapolis the police advise, "Protect your family, property, and neighborhood by turning on your front door and yard lights," and "Remember: Criminals like the dark, so make sure your yard has lots of light!"

Clearly, plenty of us have been receiving similar advice—we live in a world that is brighter than ever before, and growing brighter every year. Part of that growth comes from an ever-increasing human population, especially in urban areas. But the amount of light we are using per person is growing as well. In the UK, for example, lighting efficiency has doubled over the past fifty years—but the per capita electricity consumption for lighting increased fourfold over that time. We are choosing to light up more things, and we are lighting those things more brightly.

There's no doubt light at night can make us safer, from a lighthouse beam guiding ships from rocky coasts to simply enough sidewalk light to keep us from tripping on cracked cement. But increasing numbers of lighting engineers and lighting designers, astronomers and dark sky activists, physicians and lawyers and police now say that often the amount of light we're using—and how we're using it—goes far beyond true requirements for safety, and that when it comes to lighting, darkness, and security we tend to assume as common sense ideas that, in truth, are not so black and white.

Foremost among these assumptions is that because some light improves our safety, more light will improve our safety more. It's an assumption I will hear challenged again and again. As one lighting professional explained, "Too much light would have a negative effect, because if you look into a light, you can't see anything, you can't see beyond it." Gazing from behind his desk, he paused, "You know, a bright enough light in between us and we can't see each other—and we're sitting across from each other!"

The sky over Concord, Massachusetts, this famous town of sixteen thousand about twenty miles west of Boston, reminds me of the sky above my parents' house near Minneapolis—washed out. (Alan Lewis, whom I have come here to meet, calls it "the great yellow sky.") Of course, this wasn't always so. In 1836, for example, Ralph Waldo Emerson wrote of the stars here:

> Seen in the streets of cities, how great they are! If the stars should appear one night in a thousand years, how would men believe and adore; and preserve for many generations the remembrance of the city of God which had been shown! But every night come out these envoys of beauty, and light the universe with their admonishing smile.

This is almost like reading ancient history—stars, seen from the streets of cities? In this passage from *Nature*, Emerson looked for a way to make the point that we take nature for granted—we take life for granted—by finding an example of something so commonplace we don't even see it anymore. What better example than the brilliant starry night over a nineteenth-century Concord lit by oil lamps?

I didn't have to visit Concord to know that its sky holds many fewer of Emerson's "envoys of beauty." But I wanted to talk with Lewis, to learn more about how too much light could actually act in a negative way. A longtime optometrist and former president of the Illuminating Engineering Society of North America (IESNA), the lighting professionals who have much to say about how we light our world, Alan Lewis has spent the last forty years helping to "educate lighting people about how the visual system operates."

For example, Lewis says, most streetlights are actually designed in a way that often causes more problems than they solve.

"Badly designed street lighting, which is probably eighty percent of street lighting, are glare sources," he explains. "That is, they actually reduce the contrast of things you're trying to see rather than increase it, because of this disability glare problem that occurs due to scatter in the eye."

Disability glare from poorly designed streetlights—picture the traditional cobrahead drop-lens fixtures used on most American streets—is the main reason drivers, especially older drivers, have a tough time at night. As we age, proteins in the lens of our eye begin to accumulate, and we lose the transparency we had when we were younger. In the same way that a brand-new windshield is crystal clear but ages over time with accumulated minuscule chips and dings, these proteins reduce the eye's transparency as they scatter the light coming into the eye. The effect is that instead of going to the retina and focusing, the light is distributed across the retina, casting what Lewis calls "a veiling luminance" that significantly reduces contrast.

To optimize vision, Lewis says, the key is to maximize the contrast—the brightness difference between what you're trying to see and the background—while minimizing the amount of light going directly from the light source into the eye, because when light goes directly into the eye the greater portion of it is scattered. "You don't want bright lights coming in from anywhere but the target you're trying to see," he says. "I mean any additional source of light out there, like a streetlight shining in your eye or a headlight coming at you or glare sources on a building just makes things harder to see."

The second major factor in our seeing well at night (or not) is adaptation, the way our eyes adapt as we move from brighter areas to darker areas. Because of the way our streetlights are usually placed, our eyes constantly have to go back and forth. "If you're in a place that's relatively uniformly illuminated by streetlights, then your adaptation remains fairly constant and that's okay," Lewis explains. "But what happens is streetlights tend to get dispersed somewhat willy-nilly and so you leave this bright spot and drive into this dark spot but you're not adapted, and so visibility is actually worse than if you hadn't had the streetlight there to begin with." Lewis compares this situation to walking into a movie theater: the way it takes a few moments for your eyes to adapt. "So, as you move from lighted areas to nonlighted areas visibility can actually get worse. In many cases, an equal level of darkness is better than a sporadic light-dark, light-dark area."

It isn't only streetlights that cause this problem. The worst offenders, he says, are intensely lit places like gas stations and parking lots. About twenty years ago in America, gas stations began to increase the level of lighting, not for any real safety concerns but for marketing purposes. ("People like light, they're attracted to it. There's no question about it," he says.) "You go in and you fill up under a canopy that was highly lit from a marketing standpoint to attract you, rather than a need for vision," explains Lewis. "And then you drive out into a dark road and it may be a minute or two before you can readapt to the darkness, which can be very dangerous."

"Because you might get hit?"

"Generally you're okay," he laughs, "you're in the car. It's the other folks who have to worry."

In other words, it's for marketing purposes (to get you to stop and buy stuff) that gas stations, shopping malls, and car dealerships are lit so brightly—not, as we might think, primarily for safety. If safety were the primary goal for these establishments, Lewis and others told me, they would be lit much more dimly so that the adaptation and glare problems would be reduced. The problem is that if one business raises its lighting level, the others will feel compelled to as well because *by contrast*, their establishment will seem dim and therefore less attractive—even closed.

The same scenario holds true for our society in general. As our surroundings grow brighter, we grow used to that level of brightness, and so anything dimmer seems extraordinarily dim, even dark. This is exactly what happened as artificial lighting developed through the ages. The once glorious oil lamps became dim and disgusting with the advent of wonderful gas lighting, which then became smelly and awful and unbearably dim the moment we saw electric light. In other words, once our eyes get used to seeing brighter lights, we must have brighter lights.

Roger Narboni, a lighting designer in Paris, explained this concept to me by telling of how he'd been hired to renew the lighting in a very large, very old fish market near Paris, where the business took place between 1:00 and 3:00 a.m.

"The plan was 400 lux on the fish. But when the people selling the fish saw it, they said, This is dull, we can't see the fish. They were used to having big halogen lamps—which were hot lamps that were terrible for the fish, but they were used to them. With the new lights the atmosphere was totally different, and for them it was no good. So they said, Can you raise the level a little? And we said, Sure. And they said, Can we have double? And we said, Wow, double, okay." He laughed. So, Narboni said, they raised the level of lighting to 800 lux, but when the fishmongers came to work they asked if the lights had even been changed. "I took out my light meter and showed them: 800 lux. And they said, Are you sure it's working? Can we go higher? So we went to 1,200, then 1,600, 1,800, and they were never pleased. They kept saying, It's dull, we want more. And finally I said, Okay, forget it, because we're not going to go to 3,000 lux or 5,000 lux or to daytime. This is insane; I don't want to do that. So I quit. And I told them, Your eyes are not able to understand what's going on. And even if we put more, you cannot compare it, and you will ask for more and more and more, and it's like addicts. And they never understood that."

But a fish market in the middle of the night is one thing, I said. What about in the city itself?

"It's the same for the urban city. If you put more lights for safety, very often and very quickly people will say, Oh, we don't see enough, it's not working, people are still being attacked, and we have problems and so we should put more light. And we're going to go up and up and up. There is no limit, because the vision gets accustomed to that and we need more."

The fascinating thing, though, Narboni said, is that this works the other way, too.

"If you go to darkness, the eye opens a lot, you get more focus, and even in a very dark environment you see very well."

It's a fact most of us don't know: The human eye has an amazing ability to adapt to different lighting levels, including levels we normally think of as quite dim. While the human eye will never match those of truly nocturnal or crepuscular (active at dawn and dusk) animals, in dimly lit situations our pupil expands, our iris relaxes, and thirty times more light can enter our eye. Faced with bright lights, the pupil contracts and the iris closes down for protection. But given time to adapt to low light levels—levels that would allow the stars back into our skies and make our streets safer by eliminating glare—we actually see fairly well.

"And I try to tell the politicians, 'Try it and you'll see,'" Narboni explained. "In Berlin it's like that, it's 5 lux and everything is fine. You can see the pavement, you can see the street, and you can walk peacefully."

Then, Narboni echoed the point made by Alan Lewis: "We are mainly driven by contrast. This is the main thing in lighting. So, if you go with a high level of light the contrast will be very poor and you won't feel comfortable. And if you go with contrast you can feel safe even with darkness."

Feeling safe with darkness is difficult when we have become so accustomed to high levels of light. As Bob Mizon, coordinator of the British Astronomical Association's Campaign for Dark Skies (CfDS), explains, "We're looking at a whole generation of people—even someone my age, and I'm sixty-plus—who have grown up with lots of lights, and who have grown up with lots of *bad* lights. So people now think that not only is lighting the norm, but that really glary crap lighting is the norm."

Still, a growing number of towns and villages in the United States and Europe have been experimenting with turning off some of their lights, some of the time, in an effort to save energy—that is, to save money. And despite concern about a potential growth in crime, many of these places have experienced the opposite result. In fact, police in Bristol, England, reported a 20 percent reduction in crime, and other English towns have seen crime drop up to 50 percent since lights have been turned off after midnight. When Rockford, Illinois, decided to switch off 15 percent of its streetlights, the police chief assured the city council that no studies show correlation between lighting and crime, and that he believes lighting doesn't directly affect crime one way or the other. In Santa Rosa, California, which has decided to remove six thousand of the city's fifteen thousand streetlights and place an additional three thousand on a timer that shuts lights off from midnight to 5:30 a.m., the city hopes to save $400,000 a year. The bottom of the Street Light Reduction Program website reads, "Several academic studies have been published on the correlation between street lighting and crime. None of the studies make a direct correlation between increased street lighting and reduced crime. In fact some of the research indicates just the opposite."

Other communities have had mixed success, and not because people are fearful of saving money. After Concord turned off two-thirds of its streetlights, the outcry from the populace was horrendous, Lewis tells me and the town recently voted to turn the streetlights back on, "even though most of it is pretty bad lighting."

How does that make sense—choosing bad lighting over saving money?

"So much of it is people thinking that if there's light it must be safer. And they don't know what to look for, and they don't know what good lighting is, they don't know what bad lighting is." (Lewis later tells me, "The nice thing about educating people about bad lighting is that there's so many examples.")

"And so they just think that if you turn the lighting off crime is going to go up, or I'm not going to be as safe. And none of that actually is true, in fact in many cases street lighting makes things worse not better.

"You read the letters to the editors in the local paper, and that's what they say: You turned the lights out and now I don't feel safe walking on the street anymore, so turn the lights back on so I feel safe. Even if, by the way," Lewis says, "they never walk on the street."

I'm struck by the fact that Lewis is talking about Concord, a town with a history of famous Revolutionary War violence, but with no history of pervasive violent crime. If people don't feel safe in a place like Concord, where will they? We forget that crime tends to be concentrated in only a few places, and most places have no crime at all. This is especially true when it comes to the violent, personal crime that we tend to fear most. In Concord I found a friendly New England town, and not any place where you would expect to be attacked by a violent predator. And yet, here were glaring streetlights that did as much to impair my vision as they did to brighten my way. "They could actually reduce the lighting by about fifty percent in the downtown area," Lewis tells me, "and still have very, very good lighting."

Half a continent away from Concord, in the small northern town of Ashland, Wisconsin (population eight thousand), on Lake Superior's Chequamegon Bay, you find Green Bay Packer jerseys, Day-Glo orange hunting vests, and camouflage hats or pants or jackets almost always within view, and beer and cheese served at nearly every meal. Once a town bustling with Northwoods logging and mining and railroad activity, it has only a single ore dock remaining, unused since 1965, jutting into the bay like a broken Roman aqueduct. A natural foods co-op, a bakery, and the Black Cat coffeehouse share a single block on a single street, and some residents claim you need never go anywhere else. Except, perhaps, for a late-night drive to Tetzner's Dairy outside nearby Washburn to grab a chocolate ice cream sandwich from the freezer and leave your cash in the coffee can near the door. From the surrounding woods, on one of the nearby Apostle Islands, or better yet floating in a sailboat on the lake, the nights are still dark enough to welcome the Milky Way in brilliant detail.

But in town, light abounds. Along the lake on US Route 2 shine rows of "acorn" streetlights, the Victorian fixtures that tend to show up wherever decision makers want a nostalgic look. And in the neighborhoods you find plenty of what are, for residential and commercial buildings, the two most common sources of bright light: the "security light" and the "wallpack."

Whether in alleys, barnyards, backyards, front yards, or driveways, the white 175-watt, dusk-to-dawn security light is ubiquitous in the United States. Drive into the country and they are often the only lights you see. I remember as a child traveling south with my parents from Minneapolis to the southern Illinois farm country where my grandparents lived. If we went at Christmas we traveled long hours after dark, and I would press my face to the backseat glass, cup my eyes, and stare at the stars. Somehow, the solitary white lights that dotted the black landscape seemed part of the romance, like bits of starry sky fallen to earth.

But that romantic view hid the reality: The fact that I could see them hundreds of yards away speaks to the glare these lights cast in all directions, including far beyond the boundaries of the property for which they were meant to provide security.

During the three years I taught at a small college in Ashland, I lived within walking distance of my office and often would take the alleyway five blocks there and back, passing right under a security light. This light was ostensibly designed to illuminate a driveway and a garage basketball hoop, and I used to imagine the swish and clank and splat of a solitary sharpshooter hitting net, rim, and puddle. But I never saw this person; I only saw, from blocks away, the light casting its shadows and glare into neighboring yards and houses. Approaching the light, I would have to shield my eyes, losing whatever dark adaptation I'd gained. I never did ask the neighbors what they thought of this light. My guess? They were so used to it they no longer noticed.

That we don't notice glaring lights anymore has direct ramifications for light pollution, of course, but in terms of safety and security, because we are so used to bright lighting, we won't notice if anything out of the ordinary is taking place. In fact, we won't think to look or even want to look. And if no one is looking, lighting will do next to nothing for security.

For example, think of the many industrial warehouses spread outside every city and small town that stand unattended all weekend, every weekend. With few exceptions they will be ringed with lights, far too often wallpacks—those rectangular-shaped lighting fixtures plastered to the sides of buildings all over the country that blast horizontal light onto parking lots, plazas, courtyards...and far beyond those areas. But without a human presence—without someone watching—those lights do little more than provide illumination a criminal would need. So much so that David Crawford, founder of the International Dark-Sky Association, calls this "criminal-friendly lighting."

The joke I heard in London is that criminals actually prefer to work in well-lighted areas because they, too, feel safer. Studies bear this out: Light allows criminals to choose their victims, locate escape routes, and see their surroundings. Asked in one study what factors deterred them from targeting a house, criminals listed "belief that house is occupied," "presence of alarms or CCTV/camera outside the property," and, to a lesser extent, the "apparent strength of doors/window locks." Nowhere did they mention the presence of lighting.

"It works both ways, you see," the CfDS's Bob Mizon told me. "The people who claim benefits from lighting, they never put themselves in the mind of the criminal—what does he or possibly she need? What does a burglar need, what does a rapist or a mugger need? They need to assess the victim; they need to see what they're doing. I mean, who benefits most from a big security light at three o'clock in the morning? Is it the resident fast asleep indoors, or is it the burglar sorting his tools under the light?"

Makes sense, I said, but when I look at the webpage for the police department in the suburb where my parents live, the first thing listed under preventative measures a homeowner can take: "Home is illuminated with Exterior Lights."

The police in his town offer the same message, Mizon explained: "You must have lights to prevent crime. And when asked about the source of this information, the data, they haven't got any. They just assume it's true. The police are mired in the same degree of ignorance as society."

This doesn't mean that the Campaign for Dark Skies is against artificial light.

"It's not as though we want people stumbling about in medieval darkness," Mizon said, smiling. "I mean we don't campaign for *no* lights. That's crazy. If people want lights they should have them. It's a democracy; people fought and died for it. Let's say everybody in the village votes for street lighting. Great, they must have it. But it's got to be the right stuff. And that's what many people don't realize, that there's lighting, and then there's lighting."

Helping people understand this is a large part of Mizon's job.

"I say to them, look, there are thousands of little villages in England with no streetlights—are they hot spots of crime? No, they are not. And when you see crime on the television or you see people rioting in city centers or fighting each other on the CCTV cameras and vomiting in the gutters, are those dark places? No, they're brightly lit places. They're the brightest places in Britain, and they are the most crime-ridden places in Britain. So what's the conclusion? Does light prevent crime? Of course not, it's rubbish."

Overall, the available studies and statistics back Mizon's claim, and echo what several people told me, that the term "security lighting" is simply oxymoronic because it assumes a link between security and lighting that research does not support.

In 1977, a U.S. Department of Justice report found that "there is no statistically significant evidence that street lighting impacts the level of crime." In 1997, a U.S. National Institute of Justice report

How shielding our lights cuts out glare and improves our vision (notice the "bad guy" standing in the open gate). *(George Fleenor)*

concluded, "We may speculate that lighting is effective in some places, ineffective in others, and counterproductive in others." In 2000, the city of Chicago performed a study in which an attempt was made to "reduce crime through improved street and alley lighting." The city found that "there did not appear to be a suppression effect on crime as a result of increased alley lighting." In 2002, Australian astronomer Barry Clark conducted an exhaustive review of the research available and concluded that there is "no compelling evidence" that lighting reduces crime and, in fact, "good evidence that darkness reduces crime."

In late 2008, the Pacific Gas and Electric Company (PG&E) was required by California law to find ways to reduce energy expenditures. In an effort to look into how the reduction of street lighting might do so, company representatives asked for and received an independent review of existing research "relating to any relationship between night-time outdoor lighting and security." The review found no research that presented "sufficient evidence to demonstrate a causal link between night-time lighting and crime" and concluded: "the available results show a mixed picture of positive and negative effects of lighting on crime, most of which are not statistically significant. This suggests either that there is no link between lighting and crime, or that any link is too subtle or complex to have been evident in the data, given the limited size of the studies undertaken."

As Barry Clark argued in 2002, "Where the justification includes or implies crime prevention," lighting costs "appear to be a waste of public and private funds." Updating his review in 2011, he reiterated his earlier findings and wrote, "Given the invalidity of evidence for a beneficial effect and the clear evidence to the contrary, advocating lighting for crime prevention is like advocating use of a flammable liquid to try to put out a fire."

These studies have had little effect, however, on the perception that lighting reduces crime at night, and that more light reduces crime further. Perhaps that's because most of us have never heard of these studies, and so continue to assume a connection between darkness and crime, lighting and security. It doesn't help that a handful of studies directly or indirectly sponsored by the lighting industry or utility companies persist in claiming that lighting deters crime despite mounting evidence to the contrary. By selling more lights or selling more energy, these companies stand to gain the most wherever the lights are brightest. Widespread ignorance reinforced by questionable research has much to do with this, no doubt.

But there's something else going on here, too. You get the feeling someone who says "Then send your wife and kids into the darkness and see what happens, or ask a rape victim what they think" isn't going to be dissuaded by Clark's study or any other study. Dare to question the idea that we need lots and lots of bright lights for safety and, as Martin Morgan-Taylor of the Campaign for Dark Skies told me in London, "It will often raise quite an aggressive response from people, because it really is the fear of the dark, isn't it?"

"This most ancient of human anxieties," explains historian Roger Ekirch, "has existed from time immemorial. . . . Night was man's first necessary evil, our oldest and most haunting terror." The reasons—rational, even—that we have feared the dark of night are many: threats from wild animals, attacks from robbers or highwaymen, deadly terrain, and especially fire. Add to those reasons our propensity for irrational fears such as ghosts, witches, werewolves, and vampires and we had plenty of reason to fear the dark. Regardless of which came first—the rational or irrational—as we evolved, those fears were kept intense by the human eye's limited ability in the dark and our imagination's vivid ability to see night's demons all too well. Christianity, which saw Christ as "light eternal" and Satan as Prince of Darkness, further ingrained this view of the world. In the eyes of the church, continues Ekirch, "the devil embraced darkness, literally as well as metaphorically. Night alone magnified his powers and emboldened his spirit. Indeed, darkness had become Satan's unholy realm on earth."

But most of us no longer fear attack by wild animals, deadly terrain, or fire at night; nor do we recall our last encounter with a highwayman. And while we love them in movies, we don't normally fear meeting witches, ghosts, or werewolves in the dark—or at least, we don't admit we do.

No, now we fear each other.

Three miles northwest of downtown Winston-Salem, North Carolina, the campus of Wake Forest University, a highly ranked school that is home to more than seven thousand students, lies embraced by quiet residential neighborhoods or estates on all sides. This is where I work, where I experience night's darkness on a regular basis—leaving my office on a winter's early evening, returning after dinner to hear a visiting speaker, walking with Luna. A leafy tree canopy made of magnolias and maples, dogwood and pines, covers walkways and streets and pedestrian squares, the brick buildings made in the Georgian style. Wait Chapel rises at the north end of the main quad, spotlights on its steeple, while other lighting includes both the old and the new: security lights and wallpacks and cobraheads, but also fully shielded streetlights, "dark sky–friendly" acorns, and other attractive Victorian fixtures. A recently commissioned report on campus lighting states as a goal to "continue the intimate feel of campus."

"And that means a balance of darkness and light," says Jim Alty, vice president for facilities. "If you want to be out with your girlfriend, or if you want to talk with a colleague or classmate, you don't want to be in a brightly lit space that's making you squint. So our idea is to offer pathways that are well-lit but, once you get away from the pathways, not trying to make it look like Times Square."

Nonetheless, "for some parents we couldn't get campus bright enough," Chief of Police Regina Lawson explains. And surveys of the campus community tell her that "people are afraid, and feel that it's unsafe to walk across campus in the dark." It doesn't seem to matter that, as she says, "the reality... is that we're being robbed blind in broad daylight. The perception is that people should be afraid at night." Lawson cites what she sees as the sensationalism of today's media as heightening our perception of fear. "When I was in college you didn't have that. You felt pretty invincible, because you weren't getting that redundant, redundant, redundant replay of any acts of violence that took place."

Consider this as well: Like so many college campuses in the United States, this campus is dotted with the silver poles of a "blue light system." The idea is that if you're in danger, all you have to do is get to a tower and press a button for help. When these systems were first finding their way onto college campuses in the mid-1980s, author Katie Roiphe raised the question of their effectiveness. She argued that rather than actually providing real security to university students, these lights created a culture of fear that taught students to be anxious about darkness, about strangers, about the night. As she wrote, "red means stop, green means go, and blue means be afraid."

I wonder if she might be right. In a 2000 report titled "The Sexual Victimization of College Women," researchers at the National Institute of Justice found that "the majority of sexual victimizations, especially rapes and physically coerced sexual contact, occurred in living quarters." That is, not while the victim was walking across campus at night. "Almost 60 percent" of the rapes on campus occurred in the victim's residence, 31 percent in other living quarters on campus, and 10.3 percent in a fraternity. (Perhaps the blue light systems ought to expand their reach, with towers in fraternity and dormitory lobbies?) Campus lighting, blue light systems, statistics telling us we ought not to fear being outside at night—and still we are afraid to be on campus after dark.

"I think we're scaring the students more than we need to," says Alty. "We have some crime on campus, but not much. So why are we telling students, 'Oh, be careful, you gotta be careful'? Are we scaring them or sensitizing them? That line is blurry for me."

With rare exception, we in America are born from darkness into brightly lit rooms and grow up in brightly lit cities and suburbs, our nights both inside and out lit by electricity. By the time we reach college we already know what night is supposed to look like, and so we accept glary lights on campus because we assume they will protect us if we have to venture outside after dark. But lights alone don't protect us—being smart about our actions and aware of our surroundings does. And what an opportunity lost: Rather than college being a time when we become more appreciative of night's beauty and informed about the value of darkness, we spend four years having our previous assumptions reinforced—that night is dangerous and darkness a threat.

This is not to say that threats at night don't exist, or that we have no reason to feel anxious. It's an especially sad statement about modern Western civilization that women, in particular, are made to feel nervous being out at night.

"As a woman you're constantly looking out, trying to maintain your safety," admits Tiffany Bourelle, a professor at Arizona State University. "I don't think it's something anybody can recognize until you've been in that situation of fear."

Bourelle and her husband, Andy, have joined me on Tempe's "A" mountain to watch the moon rise over greater Phoenix. In the east, the deep dusky purple of Earth's shadow; in the west, the setting orange sun—the city's evening humming all around. The planes coming in to PHX pass directly overhead, all polished aluminum and engine whine, looking like big white flashlights strung loosely one after another in a long, loose chain stretching east until disappearing on the horizon. Except where they are broken by mountain shards rising from the desert floor, the lights sprawl in every direction to the horizon—the pink-orange of high-pressure sodium streetlights, the sharp green of stoplights, the bright white of empty parking lots. On a distant ridge, a forest of radio towers blinking red.

"If I'm alone in a parking garage late at night," Bourelle continues, "I have my keys in my hand, with my key in my index finger ready, because that would probably be my only chance. And you guys probably don't even think about that."

Listening to her, I think of Rebecca Solnit citing in *Wanderlust* "the most devastating discovery of my life": that she had "no real right to life, liberty, and the pursuit of happiness out-of-doors." She describes how in order to walk in the streets she "learned to think like prey, as have most women." She cites one poll showing two-thirds of American women are afraid to walk alone in their own neighborhoods at night, and another poll reporting half of British women afraid to go out alone after dark and 40 percent "very worried" about being raped.

Yet, as on the university campus, is the danger real or perceived? In their article "The Gendered 'Nature' of the Urban Outdoors: Women Negotiating Fear of Violence," Jennifer K. Wesely and Emily Gaarder studied "the ways that gendered constructions of public space, particularly the wilderness outdoors and urban-proximate areas, inform women's assessments of vulnerability and fear in these spaces, or their 'geography of fear.'" They found that "violence against women in the private realm far exceeds that in the public sphere," and "the vast majority of sexual abuse, rape and battering occurs behind closed doors." They argue, "Countless women are probably denied the healing benefits of wilderness because of the fear of rape behind every bush, around every corner—a fear that every woman in this culture has been taught along with Little Red Riding Hood and the Big Bad Wolf."

It doesn't seem to matter what the studies or statistics say—we remember the California college student abducted from a friend's apartment in Reno and found strangled weeks later on the edge of town, the University of North Carolina student body president kidnapped and killed. The rarity of these crimes doesn't seem to diminish our fears. We remember the sensational cases, and we are afraid. That forty thousand of us die in traffic accidents each year doesn't make us afraid of driving, but one rape or murder confirms every fear we have about night's darkness, and keeps us from going outside at night.

"And it's nothing that lights can cure," Bourelle explains. "It's because there are these stories out here, and every woman knows some woman that's been attacked, that it doesn't matter how many lights you put in a place—it's always going to be dark at night and there's always going to be shadows, and that one person in a million is going to get raped outside at night. So, there's really no way to get around it—nighttime makes you feel vulnerable."

When I hear that phrase I think of Bonnie, a friend in Albuquerque, who committed herself on New Year's Day a couple of years ago to getting out to see each full moon of the coming year. It wouldn't be enough to wave at the moon from through the kitchen window, Bonnie had to get out to an area dark enough that she could see her moon shadow. Her long-term relationship had just ended, and she was looking to reconnect with a part of herself she felt she'd given up, and to give herself a positive way of marking the passage of time rather than simply sitting at home waiting to feel better. Whether in the Sandia Mountains east of Albuquerque, the bosque of the city's North Valley, mountain biking in southern Colorado, or cross-country skiing near Santa Fe, Bonnie made a point of getting out into the dark.

"People were always asking me, 'Aren't you afraid?,' or discouraging me by saying, 'It's not safe,' or, 'What are you doing out at night?' There's this perception that the night is sooooo dangerous.

But you're much more likely to get assaulted in your home at night than you are out in the dark," she tells me.

"Women are taught to fear," she continues. "It starts with that mysterious female body stuff that guys are taught to be scared of but, really, women are taught to be scared of as well. We're taught to hide it, be ashamed of it, to not embrace it." And part of teaching women to be afraid, Bonnie says, is teaching them to fear being out at night, that bad things are going to happen. "It's much easier to control people who are scared, who won't leave the house," she argues. "It's this manufactured fear that creates a perception that something bad is going to happen to you."

The reality, she says, is that as you sit at home watching TV "something bad *is* happening—you're getting sick, and you're missing out."

Without diminishing the reality of our fears, can we ask what we might be missing? Can we ask what's lost? What do we lose—women and men alike—when we are so afraid of darkness that we never experience its beauty or understand its value for our world, while allowing our lights to grow ever brighter?

If ever-brighter lights were making us increasingly safe, that might be one thing. But as Eddie Henry, the man responsible for lighting one of London's toughest boroughs, told me, true safety "is about having the right amount of light and the right type of light and the right color of light for the right place. Rather than, we're just going to have loads of it!"

Far from being contradictory goals, lighting our nights for safety and controlling light pollution go hand in hand. In fact, one of the most compelling arguments for controlling our lighting at night is that by doing so we will actually make ourselves safer. Said another way, if we are truly concerned about the safety of our wives and daughters and mothers (husbands and sons and fathers), we will understand that light as we use it in most situations makes us less safe by impeding our vision, casting shadows where the "bad guys" can hide, and—perhaps most powerfully—creating the illusion of safety.

Lights can help make us safer, but real safety comes from being aware of our surroundings, making good choices, and not using our natural fear of the dark as an excuse to overlight our nights.

And that natural fear of the dark? I know it well.

I have been afraid of the dark since I was a child. For me it's not a fear I feel in the dark of the city, or anywhere I'm with a friend. But in summer when I get back to our lake in the northern Minnesota woods, I remember. As a child at the lake I avoided camping at all costs, slept with a night-light, and dashed home from our nearest neighbors. Even now, after all I have learned about darkness, on nights I'm alone at the lake I will sometimes stand where our gravel road begins to climb and wind through the woods behind our house and find my legs refuse to go farther. During daylight this road is easy, but on nights when my hands disappear inches from my eyes, I can't take another step.

A few years ago, I decided that if I could walk this road on a moonless night, I could overcome my fear. I knew all the rational reasons why I shouldn't fear the dark, and I figured I just had to grow up and wrestle into submission my irrational fears, which had almost nothing to do with cougars, bears, and wolves, and almost everything to do with some deranged human somehow finding his way down this single lane in the Northwoods.

But I abandoned my experiment almost immediately, and I remember the moment I did. I was standing barefoot on the dock watching a waxing moon glow over the southern bay, its light on the water the most movement around, when from far back behind the house there came an eerie, wonderful howl—a sound I had never heard at the lake before. At first, my mind thought, *coyote.* Then, just as I thought, *No, but what is it?*, a shiver slid from my scalp to my heels.

I was twenty yards from our front door, the wolf staying deep in the woods. I knew—intellectually—that I wasn't in danger, that no wolf would be interested in attacking me or anybody else. Yet the primitive fear remained—for both of us—and that gave me hope.

Let me explain.

Perhaps no animal has been more associated with devilish darkness, and dealt with more cruelly, than wolves. In western Europe, wolves have long been wiped out. And in the United States, these intelligent and social creatures have suffered a destruction that boggles the mind. From 1680—when William Wood wrote that there were far too many wolves (in New England!) to ever hope of doing anything about them—to the later twentieth century, the wolf population in the Lower 48 was reduced (trapped, shot, poisoned, stoned, burned, smoked out, drowned) to a handful of packs, cleared from nearly every state. Northern Minnesota is home to several thousand wolves, thanks to federal protection and intense human management. It's a safe bet that, without this help, wolves would have been and would still be driven even from their last refuges.

I will readily admit that I am still, especially on windy nights or nights of thunder and lightning, afraid of the dark. But I have come to realize that appreciating darkness has little to do with my conquering fear and everything to do with accepting it. In fact, my fear of the dark—or at least the response it sparks inside me—is something I value. Standing at the bottom of the road is for me the equivalent almost of standing at the open door of an airplane a mile above farm squares. It quickens my heart and makes my blood rush around. That adrenaline, that excitement, that feeling of being alive—when I feel the stop signals in my legs, it's the instinctual, natural, animal awareness in me. I don't want fear so strong that I am incapacitated. But there is fear that comes from being attentive enough that you realize there is life greater than you, life that was here before you and will be after.

That is one reason I am standing in a dark desert canyon surrounded by mountain lions; Ken Lamberton is the other. I have wanted to visit Ken in southeastern Arizona ever since reading his essay about what darkness meant to him during his twelve years in prison. I wanted to ask what it meant to have darkness taken from you, to be barred from seeing it. I wanted to talk with him about freedom, and what we lose, if anything, when we don't have access to darkness. And I wanted to know what it was like for him now to live in this dark desert canyon.

"One of the advantages of living out here is that you have this night sky," Ken says, his curly black hair and mustache silver-tinged in the moon's light. "When I was a kid I learned all the constellations and the stars. My first telescope, I must have been twelve or thirteen, I remember just looking for bright objects in the sky, and I pointed it at one, and I looked through the ocular and there was a planet with rings around it....Oh my gosh! Saturn. I was sold. Being out here now is like my childhood all over again."

We are making our way down a single-lane gravel road, pebbles kicking out before us, rolling underneath. A low waxing crescent moon rests in the western sky, casting our shadows onto rocks and prickly pear and cholla.

"When I was locked up I used to really miss the stars. On those rare occasions when there would be some kind of event and they would send us out of the cells, and you could see the stars, it was like being released, it was like being free. I mean looking up and seeing the stars, and not seeing any of that razor wire, not seeing any of those fences, and knowing that it's just light-years of space."

I don't know the whole story of why Ken was in prison. I've heard he made a mistake, and that maybe there was a judge who wanted to send a message. I know Ken's written about prison in his books. I know his wife, Karen, stayed with him the whole time, and that they have three grown daughters. I know Ken exudes kindness and care for the natural world. Before I read Ken's story, I hadn't considered the idea of having no access to darkness, of being forced to be in the light. He writes of floodlights on the prison yard creating "a hazy smog of light," and "the lights snapping on with an electric whine as dusk settled upon our cordoned backlot of desert south of Tucson."

> Extinguishing the light in my cell only allowed the bright hallway outside my windowed door to lay a column of chalky alabaster across my face. Blocking the window risked a disciplinary write-up. Covering my face invited a rude wake-up. Too many times the rapping of a flashlight on the glass disturbed my sleep as some guard on graveyard shift attempted to get me to show some skin so he knew I was real and not a stuffed blanket.

Even in the housing units, he writes, "we had no escape from the security of light."

We start our walk from Ken and Karen's small stone house, walking north past a few other houses, shielding our eyes at the one bright house, then walking without streetlights, walking with moonlight instead. Ken says, "If you want to get into some really dark walking, we could go down along this gully. We could follow an old mule road down to the bottom of the hill."

I tell him I'm all for it, as long as the lions don't care.

Earlier in the evening, Ken had shown me photos of the lions he'd photographed by remote sensor. "The first time I developed the film in Tucson, Karen went to pick up the prints and she was like, '*So where again is this exactly?*' And I said, 'I don't believe it, there's mountain lions a hundred yards from our house!' "

As we descend into the gully, the stars condense to only straight overhead as bushes and trees surround us and darkness rises on four sides. I'm trying to stay close behind Ken, the wind whistling in the juniper trees and our boots crunching gravel.

But I'm thinking, too, of Aldo Leopold's essay "Escudilla," from his book *A Sand County Almanac*, the story of a mountain not too far from here that used to be home to a grizzly—before a government trapper killed him. When I first read the essay, I was living in Albuquerque, and as soon as I could I drove down and hiked Escudilla—a country full of solitude and southern Arizona mountain beauty—and saw not a single person. Luna and my friend Rachel and I were the only ones there among the pines and aspen. And I remember thinking how different it would have been if I'd known there was a grizzly sharing that mountain.

"Yep," Ken says, "we don't see them, but they see us."

The fear that causes us to overlight the night and keeps us from appreciating darkness also keeps us from the value of fear. And just as I'm not about to bungee-jump off a canyon bridge, I'm not about to take chances in a crime-ridden part of town. But to hike in a grizzly's footsteps or a canyon with cougars, or to seek out the full moon, or to stroll through a beautiful city at night, is to know a fear that enlivens and, I think, enlightens.

In the mythology of countless cultures, the hero is called on a journey that must include an experience of a dark time or dark place. For the Greek hero Perseus that meant venturing to kill the Gorgon, Medusa, but many different stories have the same message about the value of experiencing darkness. Are we to imagine that these heroes—heroes we were to model ourselves after—felt no fear? I bet Perseus was scared, and that the same was true of other real heroes in other cultures. Because if he wasn't, why would I believe his story? Why would I follow his lead? What would I learn about real life, my life, this life now—a life that has plenty of fear? With all our lights we push away our fear, and by pushing away our fear, we are a little less alive.

"It must be a poor life," wrote Leopold, "that achieves freedom from fear."

Electric light has given us remarkable freedom, allowing us to pursue our work and play long after the sun goes down. No one doubts it brings us a degree of safety and security we wouldn't otherwise enjoy. But at what cost do we lock ourselves up behind our lights?

"We live in our own little cells, our own little prisons that we create ourselves," says Ken. Though he speaks a few feet away, I can't see his face. "When you're locked up you dream of freedom," he says. And then—and I bet he's thinking of the freedom he enjoys now, walking this wild canyon after dark—"I like the idea that it's not real wilderness unless there's something out there that can eat you."

On a dark night under a setting crescent moon I walk with fear out into that wilderness and stand still enough to just about hear my blood, ready for flight, thankful I have traveled to get here, with the stars, the lions, the darkness, a friend.

6

Body, Sleep, and Dreams

Night shift is an entirely different way of life that few people who have not experienced it can understand. We live in a state of fatigue that most people never know, or would want to know.

– Matthew Lawrence (2011)

Far from the natural darkness of any desert canyon, more Americans than ever before face the dangers of our increasing dependence on artificial light. For Matthew Lawrence and some twenty million others—a total that grows every year—the pain of working at night is a daily reality. While not all work Lawrence's 11:00 p.m.–7:00 a.m. shift, all work hours that normally find the rest of us home in bed, or at least home. It's a situation that has scientists increasingly concerned as they unravel the litany of ailments affecting those who work at night—a situation that has the potential to radically alter our relationship to artificial lighting and darkness. But while the World Health Organization's International Agency for Research on Cancer (IARC) now lists night-shift work as a probable carcinogen, and researchers have linked working at night with ailments such as diabetes, obesity, and heart problems, the truth is that nearly anyone living in the developed world is subject to the potential effects of electric lighting at night.

Having evolved over millions of years in light days and dark nights, it's only suddenly, within the past century or so, that we have disrupted this ancient rhythm. Those who work at night often do so amid constant electric light. But just by going outside at night, we expose ourselves at every turn to streetlights, parking lots, flashlights, signs. In our homes we burn our lights—including television, computer, and tablet screens—right to the moment (and for many, even after) we close our eyes.

The potential consequences of our exposure to all this light at night, scientists say, are enormous.

Take sleep, for example, or rather our lack of it. As Steven Lockley of Harvard Medical School's Division of Sleep Medicine tells me, "At the moment, we worry about diet and exercise and smoking and alcohol as risk factors for our health. As we learn more about the impact of poor sleep, it may outstrip them all." Sleep disorders are now arguably "the most prevalent health concern in the industrialized world," says the University of Arizona's Rubin Naiman, with ramifications that reach throughout society. What does sleep have to do with artificial light and darkness? Close behind the increasing recognition that "every major disease is associated to some extent with short sleep," says Lockley, is the fact that "short sleep also means long light."

Long light—electric light burning into and sometimes through our nights—is a fact of modern life, but we have only begun to make sense of its effects on human physical health. As recently as 1980, medical wisdom held that humans were immune to any effects from electric lighting. New research suggests that far from being immune to the effects of light at night (LAN), humans are highly sensitive, and that in fact when it comes to disrupting our sleep, confusing our circadian rhythms, and impeding our body's production of the darkness hormone melatonin, LAN has the power to dramatically—negatively—affect our body's ancient codes. We increasingly understand that "exposure to light at night is a completely unnatural and alien experience," explains Lockley. "And our brain thinks it's daytime, because our brain has not evolved, ever, to see substantial amounts of light at night."

Because we have changed our nights so rapidly, so recently, results from the enormous ongoing experiment we are conducting on ourselves—and most intensively on those who work the night shift—are just beginning to emerge.

During the past two decades, as the service industry has exploded in the United States and around the world, more and more people have started to work at night. Most have no choice—their employers (from restaurants to convenience stores to factories) benefit from staying open after dark, or they work in a public safety field (police, hospitals) that society requires 24/7. In developed countries around the world nearly 20 percent of the working population now works at night. While some of these workers profess to be night owls, studies show that fewer than 12 percent of night-shift workers choose to work at night because of "personal preference." Some (8 percent) choose these shifts because of "better arrangements for family or child care," and 7 percent choose the night shift for the better pay that certain professions offer their night-shift employees. But the vast majority of night-shift workers take these shifts because they have to, and thus place themselves at greater risk for physical, emotional, and mental pain and illness. At times for our safety but more often for our convenience, millions of predominantly working-class Americans pay the price for our addiction to light.

When I first spoke to Matthew Lawrence, he explained that, like many schools, Wake Forest has recently moved to a new cleaning system that asks each custodian to do one task repeatedly (as in, again and again and again) rather than be responsible for every cleaning task in a certain building. For example, an employee becomes "the vacuum specialist" or "the bathroom specialist." Lawrence says management is doing what it can to make the new roles "a profession, rather than drudgery. We're trying to make it into a real profession with real skills and documented achievements and a way to progress, to advance yourself." But he admits that custodial work is "the forgotten career, the forgotten industry." And when I ask if doing the same thing all night gets boring, he jokes, "Yeah, that's why I'm a manager." But then he pauses. "Not just boring but I would almost say soul-destroying, because you go in and put your life's effort into making an area neat and tidy, and you come back and the kids have trashed it again. And the next night they're going to trash it again. Every single day."

I'm thinking about this as I drive to campus just before 11:00 p.m. on a Thursday to meet Lawrence. Simply going to campus at this hour feels odd. Shouldn't I be going to bed right about now? On campus, that's exactly where I imagine everyone I see is headed. That is, except for Lawrence and the dozen custodians gathering in the break room. They look weary already—each shuffles into the room as though he or she has just awoken—and their shifts are just beginning. In fact, they almost certainly have just awoken after perhaps a few hours of sleep. And while Lawrence expresses satisfaction and even pleasure with working all night ("I feel like I own this campus!"), what captures my attention is his description of the physical ordeal that is night-shift work.

"I spent five years with a permanent headache," he explains. "You can be so fatigued, and you really have to learn how to manage it, even down to how to breathe. People on the day shift, they breathe all the time, and they don't even think about it. But when you've got to slug it out and keep moving all night long, you even begin to manage how you breathe and move your arms and legs. And sometimes you get so fatigued. . . . I would lie down on my bed and immediately hit REM sleep and dream psychedelic fantasies and wake up an hour later sweating and my heart pounding like

I had been for a run. That can't be good for you, right?" About his employees he says, "It's beating them up. One or two say it's the best thing, but for a lot of our people it's just very tough."

I tell Lawrence how this makes me think about my love for night, and how the night I love is a voluntary night—I get to choose when to stay up and when to sleep. But…"

He interrupts me with a chuckle. "To be bound to it with chains? It's a different story."

I hear a number of those stories as I follow Lawrence on his rounds. The first is from Joe, a veteran of third shift for the past thirteen years. When I ask how he likes it, he sighs. "Like it? It's okay. It's where the job is. I am trained in music, Christian education, and that never paid well enough. It's a mind-set, pretty much. You either fight it or you say this is what it is and you go with it. See, I work a part-time job also, in the mornings. So I go to bed around two and get up around nine. A lot of time when everyone else is enjoying a beautiful afternoon, you can't. I wake up a lot of nights before coming here and think, Oh, you gotta be kidding me."

A heavyset woman in her fifties, Sherry has done custodial work at the university for eighteen years, but has only worked the night shift for the past two. "It's been a real challenge," she admits. "It's sucked." But then she says—and I will hear this phrase often tonight, said with varying degrees of resignation—"You get used to it." When I ask what's the hardest part of the job, she doesn't hesitate. "Sleeping in the daytime is the hardest part. My sleep is broke up. Like I go home and try to sleep two or three hours, then I get up, and then I try to lay back down in the afternoon, and that's the hard part. Sundays are really hard because your whole family's together and all of sudden you have to go to bed. I look forward to Friday and Saturday night so I can get my sleep. I look forward to going to bed. It's my best sleep. A lot of people can't sleep on their nights off, but I can really wreck out."

The toughest hours of the night shift, she says, are from 2:00

a.m. to 4:00 a.m. ("Yeah, that's true," adds Lawrence. "Because even if you were partying at night, that's when the party would be over.")

"How do you get through?" I ask.

Sherry says, "Oh, you got so much work that you just can't think about it."

That may be true, but, as Charles Czeisler, professor of sleep medicine at Harvard Medical School, explains, "You can't just order people not to be exhausted when they're working at night." Unfortunately, in our 24/7 society, where airline, automobile, and train traffic continues through the night, that's exactly what's happening. Authorities increasingly cite exhaustion from working the night shift as causing—or nearly causing—catastrophic accidents.

Here are just a few examples. In 2010, an Air India flight carrying 166 passengers crashed upon landing, killing all but eight people onboard, and investigators suspect the pilot was suffering "sleep inertia" after having just awoken from a nap. In 2011, two planes heading toward Reagan National Airport in Washington, D.C., landed without air traffic controller assistance after the controller fell asleep while on duty. He'd been working his fourth consecutive night shift of 10:00 p.m. to 6:00 a.m. The same year, a tractor-trailer slammed into an Amtrak train in Nevada, killing eight passengers, and authorities suspect the truck driver had fallen asleep at the wheel. In 2009, investigators of a crash that killed ten people along I-4 in Florida found that the seventy-six-year-old driver of the tractor-trailer that slammed into several other vehicles never applied his breaks, and blamed a devastating blend of sleep loss, shift work, and sleep apnea. Estimates are that some two million Americans fall asleep while driving on the highway at night, and that 20 percent of automobile accidents occur as a result of sleepy drivers. Those "rumble strips" at the edges of the road? They really ought to be called "wake-up strips" in recognition of their primary purpose.

Headlines like "Fatigue Likely Cause of Fatal Train Crash" are becoming common enough that after a 2011 crash of a coal train left both the engineer and conductor dead, the National Transportation Safety Board (NTSB) issued a report urging the Federal Railroad Administration to take significant action to address the problem. "The human body is not designed to work irregular schedules," said NTSB chairman Deborah Hersman, "especially during the circadian trough." This "circadian trough" refers to the time between midnight and 6:00 a.m. when our bodies have the least energy and alertness—for most of us those hours from 2:00 a.m. to 4:00 a.m. that I'd heard about from the custodians at Wake Forest. Chuck, a thirty-five-year veteran locomotive engineer, explains that trains hauling hazardous material through sleeping communities are often being driven by exhausted men who haven't slept in hours. "If you find a locomotive engineer who tells you he hasn't fallen asleep on the job," Chuck says, "you'll be talking to a liar."

Exhaustion is one result of confusing our body's circadian rhythms, which evolved to the natural rhythm of bright days and dark nights. Circadian (meaning "about a day") rhythms reset approximately every twenty-four hours and control not only our sleep/wake cycle but many aspects of our physiology, behavior, and metabolism, including hormone secretion, body temperature, blood pressure, and other subtle internal rhythms. The brain synchronizes these rhythms based on signals sent by light hitting photoreceptors in the back of the eye, signals that for tens of millions of years could only mean the presence or absence of the sun, and what season it was. In short, light tells the human body to wake up, while also setting our internal clock to expect an eventual period of darkness that will signal a time to sleep. When we're exposed to electric light at night this clock is confused, with exhaustion one of the many consequences.

If you've ever pulled an all-nighter or suffered jet lag, disruption of this clock is what you're feeling. The difference between those of us who occasionally struggle through this feeling and those who regularly work the night shift is that they subject their bodies to this experience again and again, never giving their internal clock the chance to regain its natural rhythm. As if the resulting exhaustion weren't bad enough, scientists have found it only one in a long list of health problems suffered by those who work at night. "What we're also doing is messing with our internal clock organization," says Steven Lockley, who explains that each of our individual organs has its own clockwork and its own rhythm. "So that means there's a master clock in the brain, the conductor of the orchestra, if you like, and all of the different organs of the body are trying to play the same tune. They're being kept in time by the master clock, but they also keep their own rhythms to make sure the local function is correct. And so, disrupting these internal clocks is likely to be an unhealthy thing to do, because it's going to mess with how the systems have evolved to work together efficiently. And when we start to mess around with these systems, there is more risk of them going wrong."

How might they go wrong? In addition to exhaustion and its resulting increase in sleepiness-related accidents and injuries (including the crashing of trains and smashing of semis, flipping of trucks and bashing of boats, and plenty of rolling of speeding automobiles), Harvard epidemiologist Eva Schernhammer reports that "increases in cardiovascular risk, peptic ulcer disease, a higher abortion and miscarriage rate as well as lower pregnancy rates, higher rates of substance abuse and depression . . . and higher body weight due to abnormal eating habits . . . have all been reported in shift workers."

The people most at risk seem to be those who work a "rotating" schedule—the night shift sometimes, the day shift other times—rather than those who maintain a regular night-shift schedule. It's the switching back and forth between sleeping during the day and sleeping at night that prohibits the body from adapting and thus hinders the body's circadian rhythms from adjusting to a new schedule. But consider that the vast majority of those who work a steady night shift revert to a normal day/night sleep schedule on their days off, further confusing their circadian rhythms. Says Harvard's Lockley, "The clock can't adapt quickly enough. It takes about a day to shift an hour, on average. So if you go from a day shift to a night shift, it's a twelve-hour shift, and it will take you at least twelve days to fully adapt. Then when you go from a night shift back to a day shift, it takes you twelve days to go back. And of course very few people work twelve night shifts in a row. Usually they have days off, and on days off they tend to go back to what they did in the daytime." As a result, says Lockley, "Essentially, no night-shift workers are ever adapted to their night schedule."

Consequently, night-shift workers are often awake during their biological night, a time when their physiology is sending them to sleep. It's worth dwelling for a moment on what this means: This biological drive to sleep isn't something about which we have a choice, and it's not something we can overcome. We can try—guzzle gallons of coffee or energy drinks, muster a chipper attitude or brute sheer will—and it might work for a few hours. But, eventually, sleep is going to win.

And, in going to battle every night with our need for sleep, eventually our body is going to lose.

"The question 'Why do you do this work?' is kind of a moot one, isn't it?" I say to Lawrence after we've made our first stops.

"It's true, they don't see that they have an option in this economy," he says. "But most people are doing it for the sake of their family and their home life."

I hear this from several night-shift custodians, including Lawrence ("I love it," he claims of working all night. "It's the only way of life I have found so far that allows me to meet all the needs of my family"). I hear especially from women that their main reason for working at night is that they can be with their family during the day. As one woman tells me, "It plays havoc on your body. I lost thirty pounds from the start, and I didn't have it to lose. It makes your body feel run-down, tired, exhausted. But I also have a family, and so I try to get back to a regular schedule on the weekend."

"Have you adjusted?" I ask.

"I don't think I ever will."

Ironically, women who work the night shift actually have much higher work-family conflict than women who work only days. You might think that their being home during the day would help smooth things for their families, but most people working the night shift have spouses who work the day shift, which dramatically cuts down time spent together. Even when they do spend time together, the spouse working the night shift is often exhausted. Although some whom I spoke with joked about it, like the nurse in Nevada who told me her husband thought they got along better because they didn't "see each other as much," I heard the admission that the situation was "difficult at times" far more often.

"I worked first shift for sixty-five years," says a gray-haired man named Mr. Singletary. "But now I'm supposed to flip this body, make it turn into the graveyard shift—and now I know why they call it the graveyard shift. When everybody's asleep, I'm wide-awake. Eating habits, out the window. I don't have breakfast no more—when I'm getting home my wife is already gone to work." Mr. Singletary reminds me of my grandfather—it's not only his age but how he chuckles to himself after nearly every phrase. As though Lawrence and I aren't standing a few feet away, Mr. Singletary wonders aloud about how he will find time to mow the church lawn, the local football field, and his own yard.

"You're just going to have to give up sleep," I joke.

"I'm going to have to figure out something," he says, sighing. "I'm not sure how I'm going to do it." And then, more quietly, "God will show me a way out of it. He will show me a way."

Like many of the other custodians working the night shift at this well-known southeastern university, Mr. Singletary is African American. The custodian who tells me he's "used to it" because he worked from 5:00 p.m. to 5:00 a.m. for years at the local peanut factory? African American. The woman who tells me that for eighteen years she's existed on sleeping two or three hours a day? African American. The man who tells me simply, "Some people's not cut out for third shift"? African American. When I ask him what it feels like, he pauses. "You ever worked third shift before? Okay, well. Wouldn't do me no good to explain it to you, then."

Here lies another truth about night-shift work: Certain segments of our population bear its burdens more than others. Nearly 20 percent of African Americans in the United States work the night shift, for example, and more blacks work it than whites, Hispanics, Latinos, or Asians. In addition, poor and disenfranchised city neighborhoods are often brightly lit in an effort to deter crime, and poor and minority populations disproportionately fill the increasing number of third-shift jobs. As scientists affirm the connections between the flood of electric light at night and a long list of health problems, working the night shift stands to become another public health issue that certain segments of our population will deal with—and suffer from—more directly than others.

By the end of my time with Lawrence, nearly 1:30 a.m. on a day on which I woke at 7:00 a.m. and worked my usual schedule, I have grown so tired I can't concentrate on either my questions or the custodians' answers. I can't keep from yawning, either, and when I do they are jaw-stretching yawns that bring tears to my eyes. I'm reminded of a nurse I know who, when she drives home after working all night, closes her ponytail in her car's sunroof so that her head will be jerked upright if she falls asleep at the wheel.

Not to take anything away from exhaustion, obesity, diabetes, cardiovascular risk, or higher rates of abortion, miscarriage, substance abuse, and depression (to name just a few possible night-shift-sparked sufferings), but it is cancer that scares most of us most, and cancer that may finally get our attention about light at night.

An increasing number of studies over the past two decades have made a compelling case for a link between light at night and cancer, especially hormone-influenced cancers such as breast and prostate. Specifically, it seems that light at night disrupts—that is, suppresses—the body's production of melatonin, which the human body produces only in darkness, and that melatonin plays a key role in keeping these types of cancers from growing. Light from the moon, stars, candles, or fire—none of these are bright enough to cause this disruption. Only electric light does the trick.

This means, for example, that if you get up in the middle of the night and flip the bathroom light switch, no more melatonin. You may be thinking "toilet seat," but inside your brain your pineal gland is thinking "daylight!" None of the scientists I spoke with is willing to say light at night gives you cancer—it will take more studies and data, more thought and consensus. But the research seems headed in a clear direction.

The first published suggestion of a connection between light at night and cancer came from Richard Stevens in 1987. Stevens tells me he literally saw the light when he woke one night in his Richland, Washington, apartment. "I realized that I could almost read a newspaper from the streetlight coming in," he says. "And just by luck, I knew a guy in town who was doing work on light and melatonin, and then I knew another guy doing work on hormones and breast cancer in Seattle. And that was it. I asked myself, What is a hallmark of industrialization if not lighting the night?" This moment led him to develop the light-at-night theory for breast cancer, which he describes in the following way: "increasing use of electricity to light the night leads to circadian disruption which accounts for part of the breast cancer burden in the modern world and rising risk in developing countries." In turn, this theory led him to two key predictions: Because they are exposed to artificial light through the night, women who do shift work should be at higher risk, and, conversely, blind women should be at lower risk. Both predictions have since been supported.

In the early years, Stevens found more skepticism of than support for his theory, a period he describes in his article "Electric Light Causes Cancer? Surely You're Joking, Mr. Stevens," a well-told chronology of the "journey... from electric light to breast cancer." But in 2001 he was among the authors of two papers in the *Journal of the National Cancer Institute* that showed "a significantly increased risk of breast cancer in women with a history of night work," an event that Stevens calls "the turning point for the LAN/breast cancer topic."

Next, two important developments took place, one that would demonstrate the dramatic effects on tumor growth by the presence of melatonin in our blood, and another that would show the precise wavelengths of light that maximally suppress melatonin.

In the first, David Blask did key research, published in 2005, showing that human blood taken during the night in the dark (and therefore high in melatonin) slowed the growth of human tumors growing in rats, whereas blood taken during the day or at night after exposure to light (and therefore low in melatonin) did not slow the growing cancers at all. The consequences are that suppression of melatonin by exposure to light at night may then increase cancer growth rates if you already have a tumor, or may increase the risk of one developing. "This experiment," explains Stevens, "is as close as ethically possible to a direct test of whether LAN influences breast cancer growth in women."

A few years prior to Blask's work, in 2001, research spearheaded by George (Bud) Brainard of Jefferson Medical College determined the wavelength of light that most affects melatonin production in human volunteers to be blue. This finding complemented an amazing discovery, published in 2002, by David Berson and colleagues at Brown University, of a new photoreceptor cell in the ganglion cell layer of the retina—a part of the eye that was thought not to be light-sensitive—the first such discovery in 120 years. When isolated in a petri dish, this cell also most strongly responded to blue light. Because we have studied the human eye for thousands of years, we thought we knew everything about it, including how it detected light. Essentially, we believed that there was only one pathway through which light was directed to the body, and that this was through the rods and cones that give us vision. But Brainard's experiments were inconsistent with this understanding; there had to be a whole new way of detecting light for the circadian system, separate from vision. The newly discovered cells, called intrinsically photosensitive retinal ganglion cells (ipRGCs), had nothing to do with vision per se but rather were dedicated to detecting light to determine the time of day and time of year, and in the process, resetting circadian rhythms. Brainard found that while "any sort of light can suppress melatonin . . . light composed of blue wavelengths slows the release of melatonin with particular effectiveness." The new photoreceptor cell's peak sensitivity turns out to be at a wavelength of about 480 nm, which happens to be the color of a clear blue morning sky. In evolutionary terms, that our body's ability to know day from night is highly sensitive to this wavelength makes perfect sense.

The problem is that everywhere in the world—in our computer screens and tablet screens, in our indoor and outdoor lighting—we are using more and more blue light. More than 1.5 billion new computers, televisions, and cell phones were sold last year alone, and incandescent lights are being replaced by more energy-efficient, and often bluer, LEDs. "Blue" light refers to one place on the spectrum of light, and we see (or sometimes don't—think x-rays or infrared light) different colors of light because of their different spectral makeup. Unfortunately, it turns out that the wavelength of light that most directly affects our production of melatonin at night is exactly the wavelength of light that we are seeing more and more of in the modern world.

If these links prove true, the ramifications could be huge. For example, imagine if we can link the blue light of computer or television screens at night to breast cancer, the causes of which continue to befuddle scientists. Each year in the United States alone some two hundred thousand women are diagnosed with breast cancer, and some forty thousand die. Says Richard Stevens, "It could be twenty or thirty percent of breast cancer. I'm not saying it is, but it could be." George Brainard agrees: "Even if lighting is at the root of only ten percent of breast cancer cases, what we learn may help thousands and thousands of women."

While these new findings about blue light may in time lead us to change our ways, in the end, researchers caution, the root problem is not the type of light but the fact that it's there. As Steven Lockley says, "People are now concerned about the type of light, rather than being concerned about the use of light at night overall. They're missing the point. Blue LEDs or white LEDs have a place. Other lighting has a place. But all of it has to be reduced at night. They're worried that if we change to LEDs, lighting will become a problem. The problem's there already, however, because our nights are not dark."

While night-shift workers suffer the most extreme examples of circadian disruption, light exposure at night has the potential to affect everyone living in industrialized society. For example, Schernhammer found that it's not only women who work the night shift who have lower levels of melatonin but women in general (and men—other studies have linked LAN with increased rates of prostate cancer). That is, even if we're not working the night shift, we are staying up later, exposing ourselves to light at night in ways that our bodies haven't evolved to handle.

The question is how much—or little—light does it take to confuse our circadian rhythms and disrupt our production of melatonin? Are we endangering ourselves even in our homes, even in our bedrooms? Does merely sleeping with artificial light coming through the window or seeping under a closed door spell trouble? Researchers warn that while it has been shown that levels of light produced by bedside lamps, computer screens, and tablet displays are detected by the brain and suppress melatonin, the direct evidence linking light pollution and health is in its early stages.

When I ask him about this, Stevens agrees. "Before 1980 it was thought that humans were different, we were not susceptible to light at night no matter how bright, that melatonin rhythm was just fine. Then in 1980 there was a seminal paper in the journal *Science* that changed everything. But that was with very bright light. The amount of light at night that the scientific community agrees can suppress melatonin in people has been going down, down, down. But we don't know whether chronic very low light coming in from the streetlights or whatever, we don't know what effect that has."

We haven't begun to understand all of the health effects of our living amid the flood of light at night, a flood most of us are so used to we don't question it. But if we could clearly say that electric light at night gives you cancer, or at least harms you, the whole situation changes. I asked Harvard's Lockley if he thought that, based on what we know now, it was fair to think that these connections exist.

"I think that is fair. As a scientist, I can only report what is found experimentally, and those experiments have not been done. And that's why I use the terms 'possible' and 'likely.' But the multiple shift-work studies that have shown a link, coupled with the laboratory data, are good evidence to believe that there is a link even in the absence of an unequivocal clinical trial. The WHO [World Health Organization] classification as a probable carcinogen is as high as you can get without actually proving, beyond a shadow of a doubt, that shift work causes cancer."

Of course, we know some causes of cancer. We have no doubt that asbestos causes mesothelioma, for example, and that's why WHO ranks it a Type One risk. Type Two risks—the level at which the WHO has placed shift work—until recently included breathing diesel fumes or being exposed to UV light, both risks that have since been moved to Type One.

Researchers only hesitate to identify shift work as a Type One risk because there simply is not a test we can do to measure its effects absolutely. Yet we accept the connection between UV light and skin cancer enough to support a sunscreen industry worth some $650 million worldwide, even though, as Lockley says, "It would be unethical to do a study where people are purposely given UV light to see if they get cancer."

Regardless, he says, "even if I can't prove to you right now that that light through the window does you any harm, there's no need to have it. Why take the risk?"

From Interstate 694 in the outer suburbs north of St. Paul, Minnesota, I exit toward the bright red "Emergency" spelled across a dark brick hospital building. Under the parking lot's high-pressure sodium pink-orange lights a woman pushes a man in a rickety wheelchair, while two teenagers in the car next to me pump death metal out their open windows. In the waiting room, bloated blue ankles emerge from a patient's print dress, the floor shines with fresh ammonia scent, and three teenage girls giggle in the corner. At the reception desk a small blond woman answers the phones and presses blinking buttons while two enormous male security guards shift their bellies back and forth in tight black uniforms and watch for trouble.

Ever since I'd started reading studies linking night-shift nurses and breast cancer, I'd had the idea to visit an ER. I wanted to hear from nurses what it was like to work at night, and I wanted to ask if they knew about the studies. I thought it ironic that the most well-known cases of night-shift workers and the effects of light at night focused on health-care professionals. And, having spent time with campus custodians, I was curious to spend time with people who probably had more choice about working at night than they do, and who certainly were making more money.

Soon the ER doors swing open and my host, Michelle, a nurse for twenty years, takes me first to drop my jacket in the staff break room, where coffee, slushies, energy drinks, soda pop, and cookies stand ready to transport me through the night. This ER has thirteen rooms (though no bed #13) in a square, the nurses' stations and doctors' desks clustered in the middle. White coats and tennis shoes, people staring into computer screens. Near the ceiling on each wall a digital bulletin board tracks admitted patients, ranking each 1–5, with 1 meaning "you're dying," says Michelle, 2 meaning "you're really sick" ("chest pains are 2," she explains), 3 meaning "pretty typical—you need to be seen urgently," and 4 and 5 meaning broken bones and other minor wounds. The board helps everyone know what things need to be done and are being done, Michelle says. So far tonight, there are no 1s.

A forty-three-year-old mother of two, Michelle is a blond Minnesotan of Norwegian heritage. In comfortable scrubs and a light brown sweater, a necklace of ID cards, and a stethoscope around her neck, Michelle tells me she loves working at night and always has. "Before I had kids, I was a professed night owl. If I was in the middle of a book that I couldn't put down I would be up all night reading until five thirty in the morning and I would jump when the paper was delivered. I was always a night person."

The notion of the night owl is one I've heard from a number of people, but no one has professed it as freely as Michelle has. ("Not everybody has a love affair with nights like I do," she admits.) But what about this? Are some people built to be up all night? It turns out there is some truth to the notion (and its morning-loving opposite, the "lark") in that the length of the biological clock varies somewhat from person to person. For example, some people have a clock that cycles closer to 23.8 hours, while others cycle at closer to 25 hours, with the former tending to be morning types and those with a longer cycle tending to be evening types.

Age has a lot to do with this, too. One classic example: the difference between a nineteen-year-old college student's epic struggle to get to a 9:00 a.m. class, and the relative ease with which his fifty-year-old professor who's been up since 5:00 a.m. gets there. It turns out that it's actually natural for teenagers to want to go to sleep at two, three, four in the morning, which means that making them get up for school at seven in the morning is the equivalent of making a forty-year-old get up at three o'clock every day—in other words, cruel. That said, the fact someone might be a night owl doesn't make her immune from the effects of staying awake all night. "There are individual differences," says Steven Lockley, "and some adapt better than others. But virtually no shift worker is properly adapted. They might be slightly further down the continuum (of circadian disruption) than the people who are not as able to adapt to shift work, but they're still being affected."

"Tell him what you honestly think," Michelle says as she introduces me to Chris and leaves to check on patients. "I paint a pretty glorious picture of the night shift." Chris, a forty-year-old nurse in pale blue scrubs, tells me she likes working nights, too. At least, she did until her hours were changed two weeks ago, from 7:00 p.m.–3:30 a.m. to 9:00 p.m.–7:30 a.m. She's been miserable since. "The other night I thought I was going to die," she says. I tell her that's certainly how I would feel with her schedule. But isn't she used to it? "I've worked nights for twenty-one years," she says. "And I think it's just a neurotic way to live. I don't think it's normal. For instance, I have diabetes, so I probably would be in better health if I didn't work nights. My physician has certainly told me that. It's really not normal. I don't think it's very good for you. If I'm up all night, I notice my blood sugar is different. And when do you take your meds—do you take them like you're a day person or a night person? So that's been a big issue. Another thing, your motivation really goes down. I've been saying I want to go back to grad school for about ten years. I'm able to function in normal daily living—wash the dishes, clothes, drive my kids around—and I don't feel like I'm forgetful necessarily, I just don't feel like doing a lot."

"See, I don't know any different," says another nurse, Marilyn, who's come over to join us. "I started for convenience reasons, for daycare, when my kids were little. Now I just like the hours better. I have no trouble sleeping during the day. As a matter of fact at this point I sleep better during the day than I do at night. But I don't have little kids at home anymore."

"You're not worried that a four-year-old is going to burn the house down?" Chris laughs.

"No, I am not worried." Marilyn smiles. "Me and the dogs sleep all day together. I could care less. I've adjusted to it, I think. It is a very different lifestyle. People who have not done it don't understand. My kids, it's all they've ever known. To have a mom home every night and weekend would be weird to them. It's how we've always done it."

"I do it for convenience reasons," Chris says. "I'd say money is not a good motivator for night-shift people."

"Oh, no, no, no. Money has nothing to do with it," Marilyn says.

"Some people say, 'Oh well, you night people, you make a better wage than we do.' "

"That's not why we do it."

"It's not worth it," says Chris.

"It's not worth it," Marilyn echoes. "The people who do it want to work those hours. Nobody does it for the difference in pay, that's for sure. That does not make it worth it, trust me."

Chris leaves us (she's the triage nurse tonight, and it's time to update the digital bulletin board), and I ask Marilyn what *does* make working nights worth it.

"I love the atmosphere, the different personality of the people who work at night. The teamwork is better because there's fewer people here. It's more laid-back. We don't have administrators running all over. Plus," she says, "I'm in awe of people who can get up at five a.m. and go to work. I would rather stay up all night than get up at five a.m. And I also love to be home during the day and do my shopping when everybody else is at work."

Amid the various flashing lights and beeps, intercom requests, and swirl of voiced questions and requests making the ER night's soundscape (though, I notice, there is no background music—and not that I necessarily expected it, and what, after all, would it be?), an unseen woman's pathetic wail rises again and again. None of the ER nurses or doctors seems to notice. When a man shouts, "Shut the fuck up!" Marilyn pauses. "That's her husband talking to her, I believe," she says quietly.

"I mean, you do feel tired," she continues. "And a lot of people say they didn't realize how bad they felt until they got off nights. I've had people who have worked nights for a long time tell me that. You have to make yourself get up and get going. You could lie around and be tired all the time. You just make yourself plug on. I mean, I've worked only nights for twenty years, and I know I'm going to be tired because I work nights. You just get used to being tired."

I hear this often from those who profess to prefer nights, or to at least be "used to" them, But however true the night owl idea, the biological truth is that owls are owls, and humans are humans—and unlike the nocturnal bird, we have not evolved to be up all night. As Jeanne Duffy at Harvard's Division of Sleep Medicine has said, "You can't override your biology."

That does not keep us from trying, of course.

"Well, we eat terrible," says Marilyn, referring to what is one of the toughest biological challenges to working all night. "I went to the doctor the other day and they asked when did you last have a meal, and I said, 'Three a.m., I ate dinner.' "

When Marilyn goes back to work I make my way over to talk with Steve, the charge nurse, who has worked the night shift for more than thirty years and, one of the other nurses will tell me, sometimes jokes, "I'd be a hundred pounds lighter if I didn't work nights." Steve will turn sixty this summer, and, yes, Steve is a big guy.

"I do think there's an issue with weight," he says. "When I entered the field almost forty years ago I was a skinny kid. And trying to keep the weight off is pretty tough. I don't know if it's cortisol or what. I find when I work more nights I'm hungry all the time. And that's not as true when I'm working days."

Nearly every night-shift worker I talked with admitted this challenge. When it's slow you eat to stay awake, and when it's busy you don't have time to eat well, so you just grab "a bag of chips or whatever." As though orchestrated for my visit, in the break room while I talk to Steve one of the security guards is raiding the cookie tray.

Harvard's Lockley cites eating in the middle of the night as a good example of "messing with our internal clock," in this case our metabolic rhythms. "If you eat pizza at two o'clock in the morning," he says, "you're more likely to get more indigestion than if you eat the same pizza at two o'clock in the afternoon. And that's because our body clocks have not timed our digestive responses to be maximal at night; they're maximal in the daytime." So, if a person eats a meal at night, "you're eating at a time when your biology is not able to metabolize the food properly, and so you end up with chronic elevation of insulin, glucose, and fats—which are risk factors for diabetes and cardiovascular disease. And shift workers have those increased risks."

In my visits with those working the night shift nearly everyone admitted to the difficulty of fatigue, but few—even those in health care—were aware of the research indicating that fatigue was only part of the story. As the clock crept past 1:00 a.m. and I waited for Michelle to return to her desk, I thought about a discussion I'd had with a nurse from Albuquerque named Catherine. I'd asked her if she ever talked about the risks with her co-workers.

"No," she said. "Nobody talks about anything like that. In fact,

I don't have a problem telling my boss that I want a day shift whenever one comes available, but telling her that it's because I feel like I'm more at risk for breast cancer or whatever—she would look at me like, 'Huh?'

"In my profession it's an expectation that you will work nights," she continued. "Not forever, but in nursing it's just part of the culture that you work nights first. And then once you build seniority, you can apply for day positions as they come open. I never questioned the repercussions of it, because wanting to do hospital nursing, it's just something that goes along with that career choice. That's not really a justification for not being aware. But it's just part of the job."

The difference between Catherine and others I talked with is that she had made herself aware of the risks by reading some articles another nurse had sent her. A single parent in her early forties, Catherine has been struggling working at night. "For the last several months I've been having more and more difficulty, feeling more out of balance. I don't know if you've ever had the experience of having a washer go out of balance. And it goes round and round, it's just flinging around out of balance. That's how I feel a lot of time, like I'm getting swung around and never can get back into a balanced cycle. And at the same time I'm feeling really like I want a change, but I'm not able to justify the money, because it is quite a bit more money—and so feeling that I just need to suck it up. But reading the articles really gave me that extra little bit of information that helped me reach my tipping point and say, There is a reason why I'm feeling this way. There's a reason why I'm feeling depressed. There's a reason why I'm so exhausted and not feeling well many times and many days. And it's not worth the money anymore."

It's the stories of exhaustion that stay with me. Part of that is because in my visits with night-shift workers I have dipped my toe into how they must feel—my jaw-stretching yawns at the end of my custodial visit come to mind—but most of it is just hearing what these folks endure.

"What would be the best way to describe it?" says another nurse, Heather, when I ask what she means by feeling "all messed up." "It's like I'm there and I know what I'm doing, but two hours from now if I look back, I'll be like, *Gosh, did I do that?* It's like I can't remember. I feel like I'm there but I'm not completely there. Like, it's scary as hell to be driving home in the morning. You get home and you don't even remember driving home. It's like, *Yeah, that's probably not a good thing.*"

Catherine admitted to me that she carries a prescription drug to keep her awake. "That's another reason I don't like this schedule. Because I feel like I have to take things to stay awake, and then I also have to take things to stay asleep, even times when I am just completely exhausted. I think the worst time in general that's consistent for me is when I'm driving home. My drive home is about a half hour, and I feel like it's really dangerous. There are days on a regular basis where I almost fall asleep at the wheel."

Spending time in the ER makes me wonder how the night shift will change—if it will change—if more and more evidence pointing to serious health issues begins to accrue. Certainly we have a need for night-shift workers, in fact we as a society demand it. (As Michelle tells me, "You can't just say, 'Oh, you're dying of a heart attack? Sorry, the hospital closed at 10:00 p.m.' ") But how much of the risk endured by those on the night shift is simply the result of convenience or profit? How much is a result of it being the easiest way for administrators to get things done, or of outdated traditions, such as scheduling resident physicians to work thirty-hour shifts twice a week?

"We may be at the same stage now as we were in the 1950s with smoking," says Steven Lockley. "In the fifties, a few people thought smoking was bad, but there wasn't publicly available evidence at that point. It's only over the next thirty or forty years that evidence accrued to show, without doubt, that smoking causes lung cancer. Thirty years ago it would have been impossible to think of smoking being banned in public places, would have been laughed at. But that has happened. And it's happened because of the secondary effects of smoking on other people. You have the right to smoke yourself to death if you like, but you don't have the right to kill somebody else with your smoking. And so society has decided that those risks, those secondhand risks, are worth legislation and that we'll all now have to abide by them.

"The same could be true of lighting or sleep loss," he says. "The light from my neighbor's yard may cause me problems, just like one person's smoking was giving someone else lung cancer. Similarly, shift workers driving home drowsy after a night shift may fall asleep at the wheel and kill themselves—which is itself bad enough—but is completely unacceptable if they fall asleep and kill someone else also driving on the road. The effects of secondhand light or secondhand sleepiness should be considered in the same way as secondhand smoke. It's only this type of thinking that will prompt real change."

"This is what it looks like during the day in here," Michelle tells me, returning to her desk. "The light levels are the same." On cue, a nurse nearby offers a big yawn, then a quiet "uff da" like a good Minnesotan. (Wiki says, and I confirm, "Uff da is often used in the Upper Midwest as a term for sensory overload. It can be used as an expression of surprise, astonishment, exhaustion, relief and sometimes dismay.") Because I know it is, in fact, no longer day in here, I almost expect everyone to be yawning to tears. Then again, it's only 1:45, not yet between 2:00 and 4:00 a.m.—those toughest hours for staying awake—and I don't see it.

The main thing I don't see, though, is darkness, the natural darkness of the natural night. This emergency room has no windows, and there's no way to know what's happening in the night outside. It feels like a bunker, deep below the real world. And it's this artificial separation from the natural night that I'm thinking of as I leave the emergency room and drive back across the city, seeing the world differently, terribly aware of the bright interiors I'm passing and of all the many people at work inside.

Sleep. We need it like we need food and water. It's a biological need that we cannot overcome—not for long, at least. And yet many, many, many of us are not getting it: seventy million Americans suffer from disorders of sleep and wakefulness, and of these, 60 percent have a chronic disorder. When it comes to insomnia—the inability to sleep—some 20–40 percent of Americans experience it during the course of a year, one in three in a lifetime. In 2005, the National Sleep Foundation found that 75 percent of American adults experienced symptoms of sleep problems at least a few nights per week.

There is no shortage of books on sleep and sleep disorders, but few focus on the importance of darkness for good sleep, and the possible connections between short sleep and long light. Could our dozens of sleep disorders be directly related to our lack of darkness? At times it seems an obvious connection—we have all these sleep disorders and all this light—but a connection the medical sleep profession has been slow to pursue, let alone accept. Though hospitals all over the country have a "sleep center" to assist patients in solving their sleep disorders, few of these centers have been concerned with light at night.

But that may be changing. I spoke with two sleep professionals who argue that our struggle with sleep has much to do with our disregard for the dark.

"The challenge, at least for Americans, is becoming comfortable again with darkness," says Dr. Vaughn McCall, head of the Sleep Center at Wake Forest Baptist Medical Center. "We don't know what to do with ourselves when it's dark and we're awake." In a thick Carolina accent, McCall tells me he sees "lots and lots" of insomniacs, most of whom stress over the question, "Why am I awake in the middle of the night?" The understanding that this waking was probably a normal part of the human experience, McCall argues, is one we have lost with the advent of electric lighting. "When you look at diaries from the nineteenth century, you get the sense that a hundred fifty years ago, when it was dark people got in bed, and when there was light they got up. They were entrained more or less to the natural photoperiod. People might be in bed for nine or ten hours at a stretch, though not necessarily with the expectation of sleeping all that time."

In fact, that's exactly what historian Roger Ekirch found while researching *At Day's Close*—in western Europe and colonial North America before electric lighting, people went to bed when the day's light ended and got up when it returned, but without expecting to sleep throughout the night. Instead, on most nights they experienced two major intervals of sleep, called "first sleep" and "second sleep," with an hour or more of "quiet wakefulness" between. Ekirch found that "men and women referred to both intervals as if the prospect of awakening in the middle of the night was common knowledge that required no elaboration." In other words, this pattern did not create panic. Instead, people took advantage of these periods of "quiet wakefulness" to converse with their partner, make love, pursue hobbies, or even visit friends. For many, this time of waking offered freedom and opportunity the day did not allow, true especially for women who finally had time to themselves, free from the day's toils and troubles, patriarchal hierarchies, and burdens. Ekirch discovered a second major difference between modern sleepers and our ancestors: "most members of preindustrial households probably did not drift quickly to sleep. Whereas the current time for lapsing to sleep averages from ten to fifteen minutes, the normal period three hundred years ago may have been notably longer."

To my modern ears, having the time and mind-set to hang out in bed for two hours before sleeping sounds pretty great, maybe even a little decadent. A good book—even if by candlelight? A partner you love, and want to love—especially by candlelight, or maybe with a fire in the fireplace? Maybe you're sleeping outside and the night sky has your attention. But in the twenty-first century, it wouldn't be surprising if there were a medical term for this "condition," or if it were perceived as nearly un-American in its lack of productivity. Indeed, for many, being in bed but not being able to get to sleep can cause pressure to build. It doesn't help, McCall says, that sleep has become idealized, like many aspects of human behavior.

"Take sex, for example. TV and movie portrayal of sex can leave a person feeling very inadequate. 'My sex life is not like that; what does that say about me?' And with sleep, if we overidealize sleep and say, 'If you are not sleeping exactly like this, you must have something very wrong with you,' it holds the wrong standard up. We have created wrong expectations."

I sleep pretty well, I tell McCall, although I definitely feel better if I get eight hours. And I have long dreamed of an ideal schedule where I could stay up into the night until about one in the morning, savoring the quiet and solitude of the dark house, lake, or yard... and then get up at five in the morning to enjoy the dawn. I love the late night, I love the early morning, and I wish I could sleep in the afternoon.

"You should move to Spain," he says.

Exactly. And it's a great idea, except that the Spanish tradition of the siesta is melting away under the heat of expectations from worldwide capitalism that we all be open for business at all hours, and certainly over lunch and into the early afternoon. What might this world be like, I wonder, if we were moving in the opposite direction, encouraging people everywhere to take a couple of hours in the middle of the day to savor eating, making love, and sleep?

Unfortunately, that's not the way we're headed. In fact, the second thing we've done to ourselves, says McCall, is to bring electric light into our private nights, into our bedrooms. "Purely from a behavioral standpoint, that light is the apple in the Garden of Eden that leads us down the road of temptation into doing all the sleep-averse behaviors that we don't need to be involved with, whether it's watching TV or playing on the Internet. With electric lights we have the option of staying up later and compressing our time in bed. And so, suddenly it becomes abnormal to be awake in the middle of the night."

McCall suspects this "option" of staying up later because of electric lights contributes as well to serious sleep disorders such as the epidemic of obstructive sleep apnea. "Usually the primary population risk associated with sleep apnea is obesity. So to the extent that we have an obesity epidemic in the United States, we're going to have a sleep apnea epidemic. Why do we have an obesity epidemic? Lots and lots of reasons. But how does the availability of artificial light impact our eating, what and when we choose to eat? Is this any kind of factor in obesity? If you are living in a log cabin in Minnesota a hundred years ago in pitch-black dark, there's no refrigerator to raid, there's no reason to go get an ice cream and sit in front of the TV."

I'd told McCall about my family's cabin in northern Minnesota. And he's right about the connection between LAN and obesity—research on shift workers has found them at a higher risk of obesity, and a recent study on mice showed this same connection. The problem isn't being awake at night, the problem is being awake at night in the light.

For McCall, this means helping his patients "to be comfortable with being in the dark, not fretting." Because he believes one of the many solutions to insomnia is to make waking in the middle of the night a normal event, he works with his patients to help them change their thinking. "The question then becomes not so much what does this mean, but what do I do with it. This is an opportunity I've been given; what am I going to do with this?"

* * *

For the University of Arizona's Rubin Naiman, the epidemic of sleep disorders is an opportunity to revise our attitudes toward night and darkness. We meet at a restaurant in Tucson, where Naiman makes his home amid the giant saguaro cacti of the surrounding Sonoran Desert. Picture a tall-backed booth with a rustic wood table, two bowls of miso, and two bowls of stir-fry. I know him immediately from his photo—the head of white hair and white goatee give him away.

"What's interesting about darkness is that people think of darkness as being the absence of light. I think of light as being the absence of darkness," he tells me. "You can flip it both ways."

Naiman believes "our habitual use of excessive LAN is the most important overlooked factor in our contemporary sleep and dream disorders epidemic," with the result that "we suffer today from serious complications of psychospiritual night blindness—a far-reaching failure to understand the significance of night in our lives, health, and spirituality." In *Healing Night: The Science and Spirit of Sleeping, Dreaming, and Awakening*, Naiman describes his work as an attempt to restore "a sense of sacredness to our nights" and to improve our "night consciousness."

Night consciousness? It's an idea I will run into again and again: We just don't think about night that much. For example, David Crawford, founder of the International Dark-Sky Association, told me that in addition to wanting to educate "everybody in the world" about the issue of light pollution, he'd simply wanted to "get people aware again that there is a night, and that night is really beautiful and worthwhile to everybody." That basic step of encouraging people to once again be conscious of this time and place where we spend half our life has led Rubin Naiman to a career that has included working through dreams of death and dying with cancer patients and helping soldiers who are suffering nightmares after returning from war in Afghanistan and Iraq. He's frustrated, he says, by the traditional field of sleep medicine, which he sees as tightly framing night, sleep, and dreams as strictly objective and scientific phenomena, and draining these experiences of anything personal or subjective, let alone sacred or spiritual.

"On a philosophical level, sleep disorders make sense," he argues. "We so discriminate against night. We repress it and then push it away. Part of that is a denial that there is anything there worthwhile. For example, a lot of scientists are trying to figure out why we sleep so they can do away with it. It's pesky. We just have to learn how to recharge those batteries."

I'm reminded of the character Seven of Nine from one of the old *Star Trek* series, half-Borg and half-human, and one image of what we imagine sleep could be like in the future: She didn't sleep—she went to Cargo Bay Two and stood inside an energy pod where this green electrical energy flowed into her neck. "She was recharging her battery." Naiman laughs. "There was nothing personal about it. It's a very mechanistic, soulless view of human beings. The presumption is there's nothing down there in night or in darkness that's worth seeing."

The truth, he explains, is that there is dream material; there's a level of surrendering that we don't understand. "If you just consider the possibility that there's something there, then it's really interesting to let go and to sleep. But most people when they descend into the waters of sleep, instead of aiming at the depth of that, they've got their sights set on the morning shore of waking. They really aren't going *to sleep.* It's as if this were an overnight mystery tour, but they are already thinking about where they're going to be the next day."

Naiman says that when he talks with people about this, they begin to consider that maybe sleep isn't just eight hours of being turned off, and to consider instead that they can have a relationship with night, "with all of the demons and angels, all of the qualities that lurk there."

Some of that can happen if people are willing to be touched by nature, Naiman explains, and cites as one example Thoreau's fishing by moonlight in *Walden*. "There's a great scene where he's on the pond and the stars are reflected. And he doesn't know if up is down, if down is up."

I know the scene. Thoreau has drifted off into philosophical reverie when suddenly a fish tugs at his line:

> It was very queer, especially on dark nights, when your thoughts had wandered to vast and cosmogonal themes in other spheres, to feel this faint jerk, which came to interrupt your dreams and link you to Nature again. It seemed as if I might next cast my line upward into the air, as well as downward into this element, which was scarcely more dense. Thus I caught two fishes as it were with one hook.

"It's a beautiful scene," Naiman says, "but it's also a shift in consciousness. So you can get that from nature, and you can also get that internally, I think, when you are willing to consider that there's something in there."

I admire Naiman's insistence that night has its own qualities, that it is distinct from day and not simply day without light.

He nods. "There's this notion that anything alive is in motion, and that's not true. I walk up and down this hill near my house every day and the saguaros are always perfectly still. But when you visit this same place over and over, through spring and summer and fall in the morning, winter in the late afternoon, you see the motion in stillness, the animation, because you see it in a dance with the light and the clouds. It's alive, it moves in a different way. Similarly, I think it's a question of seeing the life in night itself. To do so you have to meet it on its own terms, and you have to get real still to hear it, to feel it, to sense that it's alive."

5

The Ecology of Darkness

Humans are animals as well, and there's no reason to give ourselves any higher level in the ranking than everything else. And so when light/dark cycles mess up seasonal patterns of trees or breeding cycles of amphibians, which I think is quite well established, there's no reason to think it's not doing the same to us.

— Steven Lockley (2011)

The Massachusetts woods where Henry David Thoreau lived from mid-1845 to mid-1847, gathering impressions for *Walden*, were clear cut soon after he left, his one-room cabin sold to farmers who used it for storing grain and then dismantled it for firewood—an inauspicious start for an eventual National Historic Landmark. Thankfully, the woods have long since grown back and the cabin site is preserved within Walden Pond State Reservation. And tonight, I can't wait to see it after dark. At dusk, I park in a shopping center lot at the edge of town, tuck past a restaurant, hop the railroad fence, and start walking the tracks that lead to the pond.

I think Henry would be proud. "It was very pleasant, when I staid late in town," he wrote of Concord, "to launch myself into the night, especially if it was dark and tempestuous, and set sail from some bright village parlor or lecture room or shopping center parking lot." That last part isn't true, but I imagine myself in his footsteps as I walk one tie at a time toward his woods. Certainly there's a sense of solitude—and he wrote a whole chapter on that. Every house I pass sits turned away from the track, with aluminum and wood fences keeping railway from backyard. Peering down the tracks until they curve away, looking back more often than I need, I'm alone but for the blinking glowing lights of hundreds of green-yellow fireflies, rising from trackside bushes in bobbing, floating flight.

How different "darkness" was in the mid-1840s when Thoreau would have followed similar tracks. Gas lamps had only recently arrived in Boston, twenty long miles away, and wouldn't have affected night here; Concord had few lamps brighter than those fueled by the oil from harpooned whales. "I have heard of many going astray even in the village streets," he wrote of his hometown, "when the darkness was so thick that you could cut it with a knife, as the saying is." I know I won't see what he saw, that I should expect instead the "great yellow sky" looking east toward Boston, and that only now, as I leave the last of the Concord lights, are my eyes beginning to transition to night vision. But though my map and scale assure me I'll be lucky to find a sky anywhere close to Bortle 5, I wonder if the woods around the pond might tell a different story.

Thoreau had the woods to himself after dark, writing, "at night there was never a traveller passed my house" and "the black kernel of the night was never profaned by any human neighborhood." The reason? Same as it is today: "I believe that men are generally still a little afraid of the dark," he explained, "though the witches are all hung, and Christianity and candles have been introduced." True, true, and true, but I still have to laugh a little when the hair on the back of my neck rises as I pass under Highway 3 and take a path into the woods without a thought aside from *I sure hope this is the way.*

The common perception of Thoreau is that he went into the woods, lived in a cabin, and... end of story. But he never entirely left town life, strolling into Concord regularly to dine, pick up supplies, and have his mother do his laundry. Some find that these facts taint his purity as a wilderness saint—that if Thoreau truly wanted to be like Thoreau, he would've stomped off into nowhere and survived on his own. But Thoreau wasn't escaping civilization as much as removing himself in order to gain perspective. He readily admits he always intended to return, and after two years, two months, and two days he did. *Walden* is as much about living in civilization as it is about living in the woods, maybe even more so.

I have always thought that when Thoreau wrote that he "went to the woods" because he wished "to live deliberately" a big part of that was to live in a way that increased his awareness of and sensitivity to the world—to both the human and nonhuman nature—around him. America in the 1840s was in some ways a lot like it is today: fast and getting faster, loud and getting louder, new technology everywhere changing people's basic understanding of daily life. And, I would say, no time for living deliberately.

One of the best-worst things about going outside at night to soak up the stars and sounds and scents is that the night has its own time schedule. The moon rises when it rises, shooting stars never announce their shooting—even the sounds and scents come when they want; you can't just order them up. Thoreau went to the woods to get away from crazy-making speed and noise and new invention in order to have time to gain the awareness and perspective he desired, the fruits of which ("what it had to teach") he could bring back to Concord. It's the classic hero myth across cultures, across times: The individual goes on a journey, experiences challenges (which always include a dark place or time), and returns home with wisdom and/or riches.

"It is darker in the woods, even in common nights, than most suppose," Thoreau advised, and he's right—I soon have that Blair Witch sense I'm tromping in circles. On the train tracks there was still enough light—sky glow from Boston and all points east—to see a ways around. But in the woods the dark rushes right up and my headlamp casts a sort of surreal blood-reddish light on the dead leaves at my feet. I am grateful for that light, an accessory Thoreau did not enjoy. He writes that he "frequently had to look up at the opening between the trees above the path in order to learn my route, and... to feel with my feet the faint track which I had worn." So I've got him beat there. But he's got me beat in that I don't really know where I'm going, and my sense of direction feels skewed. I keep thinking the pond has to be just down the next slope or just around the bend. And when the path starts to drop and curve, I think, Yes! But no. And—here's the thing—if it were daylight I still wouldn't know my direction, but I wouldn't be on edge. I wouldn't be imagining a stranger suddenly standing at the side of the path. I wouldn't be feeling myself stumbling, a branch reaching out to claw and scratch. I wouldn't be covered in sweat, and just on the verge of thinking, *Stupid, dumb, why...?*

Then I hear the frogs. Singing summer frogs, savoring June dark.

And that is how I find Walden Pond at night, following their sound.

At pond's edge I crouch by smooth black water. To the east across the pond the only lights are a single "security light" by the visitors center and the occasional car—the hiss-hum of tires on asphalt, the far-off flash of twin white beams—but the Boston-lit sky rising yellow-white behind makes the pond's night brighter than the night I've just come through in the trees. The gurgles, burps, and pops of a summer freshwater pond, a dog's distant bark, a bat flickering by. I hear an owl and nod. "I rejoice that there are owls," wrote Thoreau. "They represent the stark twilight and unsatisfied thoughts which all have."

Thoreau clearly had plenty to say about darkness, about experiencing night in these woods. In his posthumously published essay "Night and Moonlight," he wrote, "Is not the midnight like Central Africa to most of us? Are we not tempted to explore it?" He's said to have been planning a book just before he died in which he would have done just that. He would have had Class 1 skies at the pond, Class 2 in town, and I wonder what he would have found. I hold my hands up to block the Boston sky, I close my eyes—I wish I could know here the night he knew.

And still—the rush of feeling that was hearing the frogs then sensing the pond then striding down to its edge—to be here at night, alone, knowing that Thoreau spent his nights here alone—I won't say I sense his ghost, but to stand at pond's edge and step back up into the woods to his cabin site, its small (ten-by-fifteen-foot) foundation footprint marked by short stone pillars, is to imagine his presence. Over a century and a half this site has seen thousands and thousands of day visitors but has known far, far fewer at night.

To be alone in the dark is to drop back through the years. I imagine him sitting here by himself (as I am now) thinking, "This is a delicious evening, when the whole body is one sense, and imbibes delight through every pore. I go and come with a strange liberty in Nature, a part of herself."

Thoreau's writings are a font of memorable sayings, perhaps

Thoreau's cabin site, marked by stone pillars and facing Walden Pond. *(Paul Bogard)*

none better known than "In Wildness is the Preservation of the World," from his essay "Walking." The aphorism is often mistakenly repeated with "wilderness" in place of "wildness," much as Paris is often referred to as the City of Lights. And, as with the dueling descriptions of the French capital, we might be tempted to ask the difference. But "wildness" makes the phrase infinitely more powerful. While we usually think of wilderness as a particular place or type of place, wildness is a quality we can find anywhere (as indeed Thoreau found it—or found it lacking—in certain books, animals, and people). It makes sense we would find wildness in wilderness, for example, but we could also find it in cities, in our thoughts and choices, in our daily domestic lives. The history of Western civilization is full of attempts to stamp out wildness—the unknown, the mysterious, the creative, the feminine, the animal, the dark. Thoreau saw an American society hell-bent on fencing in, wiping out, using up, trampling down, or blocking every trace of wildness, and declared it antithetical to sustainable life.

But if wildness preserves the world, what preserves wildness? In his chapter from *Walden* titled "Solitude," Thoreau hinted toward a possible answer.

> Though it is now dark, the wind still blows and roars in the wood, the waves still dash, and some creatures lull the rest with their notes. The repose is never complete. The wildest animals do not repose, but seek their prey now; the fox, and skunk, and rabbit, now roam the fields and woods without fear. They are Nature's watchmen,—links which connect the days of animated life.

Across the country from Walden and more than one hundred fifty years after Thoreau, Travis Longcore and Catherine Rich have created the Urban Wildlands Group in Los Angeles. Dedicated to "the conservation of species, habitats, and ecological processes in urban and urbanizing areas," Urban Wildlands ranks light at night high on its list of concerns.

"Back in 2002," Longcore tells me, "if you had Googled 'impact of light on wildlife,' or 'wildlife and night lighting,' you basically got nothing except maybe birds and sea turtles." In 2006, Longcore and Rich set about to change that by editing *Ecological Consequences of Artificial Night Lighting*, a collection that gathered the current research on light pollution and ecology. Along with articles on birds and sea turtles, the pair added pieces on bats, moths, fireflies, reptiles, amphibians, salamanders, fishes, and seabirds. Even so, most striking about this impressive anthology is that it remains remarkably thin—their section on "Mammals" has two papers, for example; their sections on "Fishes" and "Plants," one each. For although, as they write, "natural patterns of darkness are as important as the light of day to the functioning of ecosystems... as a whole, professional conservationists have yet to recognize the implications."

With at least 30 percent of all vertebrates and more than 60 percent of all invertebrates worldwide nocturnal, and with many of the rest crepuscular, those implications are enormous. While most of us are inside and asleep, outside the night world is wide awake with matings, migrations, pollinations, and feeding—in short, the basic happenings that keep world biodiversity alive. Light pollution threatens this biodiversity by forcing sudden change on habits and patterns that have evolved to depend on light in the day and darkness at night. (In just one example, circadian photoreceptors—those same ones on which we humans rely—have been present in the vertebrate retina for at least five hundred million years.) Aside from crazy-looking ancient fish and their bottom-of-the-ocean ecosystems (and those of caves or soil), every creature on this planet has evolved in bright days and dark nights. None has had the evolutionary time to adapt to the blitzkrieg of artificial light.

Significantly, Longcore and Rich make a distinction between "astronomical light pollution" and "ecological light pollution." They define the latter as "artificial light that alters the natural patterns of light and dark in ecosystems." Longcore says, "We had to do that, because the idea of 'light pollution' is very much an astronomy/astronomer focus. You could have a dark sky–compliant light—pointed down—and still do a lot of damage."

Light at night impacts wildlife in five primary areas: orientation, predation, competition, reproduction, and circadian rhythms. If we have heard anything about light at night and wildlife, we've probably heard about orientation. This is the problem of insects drawn to streetlights, of migrating birds being attracted to lit-up city buildings or communication tower lights, or of beach-born sea turtles heading the wrong way—toward streetlights and hotel signs—and ending up crushed by truck tires or plucked easily by predators.

Introduce artificial light to a dark environment several billion years in the making and suddenly some species find themselves exposed to increased predation—and so reduce their foraging time. Introduced light means new pressures of competition between species, some of which adapt better than others. Artificial lights confuse reproduction cycles—the flashing attraction signs of fireflies, for example—or the internal circadian rhythms that synchronize the internal processes of birds, fish, insects, and plants, just as they do for humans. Beyond individual species, these light/dark rhythms also shape seasonal changes such as migration. Entire ecosystems shift as the length of light changes through spring, summer, fall, and winter. As one biologist told me, "We have levels of light hundreds and thousands of times higher than the natural lamp during the night. What would happen if we modified the day and lowered the light a hundred or a thousand times? Of course, the damage would be much worse. But it is a metric. You cannot modify half the time without consequences."

Some may ask why any of this matters for humans. But when we talk about ecological light pollution, we talk about the health of ecosystems, and no matter who we are or where we live, we live as part of one. Our ecological knowledge is really knowledge of our own health.

"I've taught big intro classes where I would say, 'Raise your hand if you can name three kinds of breakfast cereal.' Every hand goes up," Longcore says. "'Raise your hand if you can name three TV sitcoms.' Class of two hundred, every hand goes up. 'Name three species of birds on campus.' Uh... the black one? 'Name three plants.' Um... grass?" Longcore laughs. "I'm not making fun of these guys—this is the way we grow up in this country. We're all urban. If somebody does have answers to those questions, they've grown up rural. People who grew up in a city have never needed this knowledge."

For Longcore, that bodes ill for the future. In his work with Urban Wildlands in Los Angeles, a city he calls "the poster child for light pollution," he includes darkness as part of a larger ecological knowledge. "Unless we pay attention to nature in the places where people live, there will be no constituency for nature in the places where people don't live," he says. "People who grow up and never have the opportunity to go to a vacant lot and play with woolly bear caterpillars or raise a swallowtail butterfly from a larva or see the Milky Way just cannot have the sort of deep connection to land and nature on which our whole conservation enterprise in this country is based."

Every year of my life, I have come to this lake in northern Minnesota. In my earliest memories I am standing with my father on the dock watching the slow, straight line of a satellite through a sugary spread of stars; I am lying on the crunch-white snow-frozen lake watching the moon through my new handheld telescope. When I was a child in the seventies, night here would probably have ranked a 2 on Bortle's scale, a "truly dark site." I certainly remember stars thicker thirty years ago than now. I remember that really until the last ten years or so. While our darkest nights might still rank a 3, the local lights from Brainard, Longville, and even tiny Remer are pushing us toward Bortle Class 4. I hate to imagine where we may be headed.

Still, this is the night I know best. From the end of our dock, from the edge of the woods, from our screened-in porch, I watch and listen, and on nights when the lake calms, I pull the ancient aluminum canoe from under the cedar trees and push it into water heavy like black oil, though clean and clear and cool. I back from dark shoreline shadows, paddling through stars, and raise a gold moon from the trees. Here the moon is as it ought to be everywhere: big, bright, beautiful. It moves through its seven phases (eight if you count "new") with confidence, climbing from the woods behind our house and crossing over the lake at its own quiet pace. Its light is a gift from the sun, a reflection of the star shining on the other side of the world. Remarkably, the moon's gray ash and rock reflect only 7 percent of the light sent its way, about as much as a sidewalk. But that's enough light to illuminate the forest, bringing it alive with the scents and sounds of countless crepuscular and nocturnal species flying, hunting, singing, breathing. In ecosystems all over the world it's the same: the dark provides cover, the moon provides light, and while the humans are home in our boxes watching our boxes, the nocturnal creatures keep this world alive.

On the shore, on the dock, paddling on clear water—the lake is where I learned. Waiting for the moon, then getting out into the night.

The canoe gleams—I could read in this light. I'm glad to have my Twins cap, pulling it down to shield my eyes as I paddle to the middle of the lake. Perseus emerges over the horizon, the Summer Triangle overhead. Lying back, I let the canoe rotate away from the moon. A splash startles me, a fish for sure. Then silent and still again. As I drift over the sandbar something scrapes the bottom of the canoe. "Weeeeeeeeds!" we would cry as kids. A freaky sound, and I nudge the canoe into deeper water. Everything very quiet, very still. And then the sound of a truck on Highway 6, a mile away, the sound as it floats around the flat water amplified, for miles coming and miles going, sputtering and puttering and passing its gas from one horizon to another.

From shore, the barred owl's hoo-hoo, hoo-hooooo, in the water, frog songs and fish jumps. The lake pulses with life. I hear bubbles that have risen from bottom weeds burst, imagine walleyes and northerns cruising beneath the canoe. A loon calls, a plaintive wail so mournful that if you didn't know better you might go for help.

We swam with flashlights one summer as kids, my cousins and I. Straight beams of white light swording through black water. We had heard the story of two friends snowmobiling the winter before, of their going through the ice, drowning. I used to envision my beam finding skeletons still strapped to swallowed machines.

Jorge Luis Borges wrote, "I think one should work into a story the idea of not being sure of all things, because that's the way reality is." There is so much unknown in the lake's wild night. The way owls hunt and fly without sound in the dark. The way wolves drift through the woods like smoke, evaporating at the first hint of morning light.

"Yeeeeeeeeeeeeeeeeaaaaaawwwww." The owl behind our house releases this long yowl, a descending tremolo at the end, then launches into several hoo-hoo, hoo-hooooooooos, and is answered by a deeper-voiced owl from down the shoreline.

I sit listening to these two owls going back and forth behind me as though they will never stop. Then, this is what I do: With two strokes I turn the canoe toward shore and set the paddle down. Two strokes as quietly as I can, and the owls stop. *They hear me*, back in the woods on their branches in moonlight.

"You can hear better in the dark," says Joseph Bruchac, a Native American writer and teacher, when I ask about our senses at night.

"A Cherokee friend of mine told me about something he was taught as a child, part of traditional Cherokee culture, called 'opening the night.' What you do is you go out into a dark place, it could be on your back porch in the old days, and you sit, and you listen to what's around you in a very close circle, within arm's length. You concentrate on that, then you double the circle, and hear everything beyond there. And you keep doubling that circle. And he said it would reach the point where at night you could sit down and hear things a mile away."

Talking with Bruchac, I think of John Himmelman telling me he likes to "tune in to" one particular cricket or katydid ("leaves-with-legs," he describes them), like singling out one particular instrument in an orchestra. Author of *Cricket Radio: Tuning in the Night-Singing Insects* and several other books, Himmelman lives on five wooded acres and says, "I like being able to pick out the individual callers. It takes me away, extending my consciousness beyond where I am. Almost like I'm taking a tour of the night forest."

Himmelman says he notices the sounds of crickets...and of katydids, of singing birds and chorusing frogs, even of "caterpillar-droppings bouncing off leaves in the summer, and the rustling in the understory along woodland edges...because I have consciously tuned in to them...since I was a very young child." He often leads tours to listen to night-singing insects and says people are delighted by what they hear. "They're so used to being inside, it's already an adventure when they show up. We don't even get started until after nine p.m.," he explains, and he says that the adults are as excited as the kids. "The most common thing I hear is, 'I never realized these were out here!' If I only find one or two things, that's a slow night for me. But for these people who have never really gone on a night walk, they're thrilled."

Walking on Cape Cod in the 1920s, the American writer Henry Beston mused on "those trillions of unaccountable lives, those crawling, buzzing, intense presences":

> It occurs to me that we are not sufficiently grateful for the great symphony of natural sound which insects add to the natural scene; indeed, we take it so much as a matter of course that it does not stir our fully conscious attention. But all those little fiddles in the grass, all those cricket pipes, those delicate flutes, are they not lovely beyond words when heard in midsummer on a moonlight night?

Himmelman tells me that insect song has been around forever; that fossils of sound-producing Ensifera (katydid and cricket) forebears date back two hundred fifty million years. As such, it's a sound humans have always known as part of their nights, a sound Himmelman explains as that of "two hardened wings being rubbed together, multiplied many times over." Crickets and katydids sing at night to avoid daytime predators, and sing in chorus to disguise their location from nighttime foes. "A singing insect is compelled to sing," Himmelman writes. "To silence it, you'd have to tie its wings behind its back." But instead, he suggests, think of these sounds as "a gift we humans have been given. We can take pleasure in hearing such sounds. It doesn't always matter why that sound is being made, or who or what is making it. Sounds can enrich our lives."

The same is true of scents, and the night air is rich. During the day, rising warm air carries the earth's scents away, but as night temperatures cool and night winds calm, those scents stay close to the ground, waiting like messages for those creatures who can receive them. Pollinators, scavengers, hunters, and prey—the scents from the earth are maps of their nocturnal world, directing travel, pairing species. Saying, *Come get me.* Warning, *Stay away.* Even for we olfactory-limited humans, night air holds overwhelming power, with scents that can send us instantly across countries and oceans, and back through time.

In the house where I grew up—with my room on the second

floor, in the northwest corner—I set my bed near the windows so that, with the lights out, I could lean toward the screens, close my eyes, and imagine the lake, or my grandparents' house, or even some unreachable place from the past or the future or inside myself—I didn't know, but I wanted to be there.

Fall nights spoke of fires and wood smoke, my parents' black Franklin stove burning the split lake logs of birch and oak, the smoke of summer infused in autumn, drifting through the neighborhood. In winter, the air told of snow and cold already in forests and lakes across the border, too soon coming our way. In earliest spring I would open the window an inch and breathe a fresh wet chill, the coming season somewhere south, red cardinals and rust-breasted robins moving their songs toward us some each day. I would press my nose near and could barely stand it—the scent so full of intoxicating promise I would have to pull away.

For moths, the night's scents mean everything—life instructions, keys to mating, perhaps even their muse. Mostly we mock their attraction to flame, to fire, to electric light. Mostly we know them as we know any of night's wildlife—dead, at the base of our light fixtures, under a rolled magazine. That we can be so quick to kill moths is ironic, for they do so much to keep our world alive. While each individual moth lives only a week or two, collectively they help to pollinate some 80 percent of the world's flora. Estimates of moth species range from 150,000 to 250,000 worldwide, so no matter where we live they share our lives.

While some moth species can have serious costs for agriculture, only 1 percent of moth species give the rest a bad name. The others are harmless. But because they come out at night, we ignore them, fear them, squish them without thought. And in their attraction to light they suffer tremendous losses, night after night. In addition to their irreplaceable role as pollinators for the flowers, bushes, cacti, plants, and trees we benefit from and enjoy, moths are a vital part of ecosystem food chains. While those species that eat moths enjoy a momentary benefit from moths being attracted to light—the Luxor beam comes to mind—those lights actually act as a vacuum, sucking such protein from the ecosystem, the food and fuel on which creatures further up the food chain rely.

Ecologists are only beginning to understand the long-term impact of this phenomenon, but Himmelman calls a world without moths "bleak," and not simply because of their economic or utilitarian value. Of the luna moth, he writes:

> Its beauty is ethereal and its nature ephemeral; it doesn't live much past a week or two. It emerges from its cocoon in the leaf litter, then mates and lays eggs within the first forty-eight hours. Then it has the rest of its nights with nothing to do. No purpose. It doesn't eat. It doesn't drink. All its energy comes from the leaves it ate as a caterpillar. That energy is finite and cannot be replenished. It is like a toy airplane powered by a twisted rubber band. Once the rubber band unwinds, the propeller stops turning and the plane falls to the ground.

Because this happens at night, most of us never see. This moth, with its colored wings and long tail, a nocturnal butterfly in the light of the moon, has no purpose as it shares our world while its energy runs dry. No purpose, you might say, except for beauty. If their beauty was merely utilitarian, you would think after copulating they would just die. But instead they stick around for a few nights. I like the thought that this nocturnal butterflylike creature is out at night fluttering around for no apparent reason except to make this world more beautiful.

Once at the lake I found a small brown moth in my bedroom, and my first thought was to squish it. But instead I stopped and watched, four wings, two on each side, tiny eyes and antennas. A friend who studies moths describes these small moths with their sloped wings as "fighter jets," this one resting on the tarmac of my bedroom door. And I found that this tiny flier can do things fighter jets can't. When I cupped the moth to my chest and released it into the night, it rose to reveal an autumn-orange underside. It was a member of the Catocalas, or underwings, group, and—as other moths have evolved "eyes" on their wings to mimic predators and bright wing colors to say *Toxic!*—this underwing color likely evolved to startle predators. But instead it startled me, a hidden flame in the lake's wild night.

I remember one summer when the local news was of a collision between a timber wolf and a motorcyclist. Both were dead, the sixty-two-year-old man thrown into a roadside ditch, and the wolf of unknown age found under the bike. They must both have been surprised, the meeting taking place before dawn, not far from the lake. I might have heard the wolf the night before, might have heard the motorcycle as it growled by. I don't know if the man was wearing a helmet. The wolf was wearing its summer coat, and perhaps, having just stepped onto the highway, had paused without intending to. Some scent caught its nose, or some sight caught its eye—the rose pulse of the waning crescent moon just above the horizon, perhaps—and so this wolf paused on the highway when, on any other morning, he would have continued trotting across the blacktop, the broken yellow line, disappearing back into pines and oaks, the woods called home. And the man on the machine would have come flying around the highway's curve in time to see the gray hind legs, the great silver coat, the glowing amber eyes. Instead, maybe the man was looking at the moon, too, wondering why it was still up in the morning. Maybe that was his last thought. Or maybe that last thought was, *My gosh, what beauty.*

Every day in the United States alone more than a million birds and animals die on the nation's roads and highways, and because so many species are nocturnal or crepuscular, more than half of this carnage occurs at night. These nighttime collisions are incredibly costly to humans as well. In fact, at least statistically, deer are far more dangerous than mountain lions or bears or, certainly, wolves. Every year in the United States more than two hundred people are killed in deer-vehicle collisions, the most dramatic result of the more than one million annual deer-vehicle collisions that cause ten thousand personal injuries and cost more than $1 billion in damages (and, of course, the deer do not fare well, either). Studies show that increasing highway lighting to reduce such collisions is ineffective, and actually makes it far more difficult for wildlife active at night, at dusk and dawn, to avoid collisions. Animal eyes that see so well in darkness or dim light—blessed with more rods than cones—are blinded by our headlights and streetlights. "On the issue of designing highway lighting to minimize road kill mortality," writes Paul Beier in *Ecological Consequences of Artificial Night Lighting*, "our knowledge of mammalian vision is sufficient to conclude that, from the animal's perspective, less is better."

Less light on our roads, fewer streetlights and not so bright, is not only better for wildlife but safer for us—we drive slower and pay more attention when we have to rely on our headlights.

But how to do with less light on our streets and highways? One innovative concept comes from a design cooperative in San Francisco called Civil Twilight. Their award-winning idea: streetlights that respond to moonlight, or "lunar-resonant streetlights." Relying on LEDs and highly sensitive photosensors, they would allow the level of brightness from the streetlight to balance the level of light from the moon. On nights when there is no moon, or just a crescent, the streetlights would provide enough light for pedestrians and drivers. On full-moon nights, the streetlights would dim to barely on. Civil Twilight estimates that their idea could save more than three-quarters of the money spent on streetlights, as well as bring the ambience of moonlight back to our streets.

The idea borrows from an old one. In the eighteenth and nineteenth centuries, street lighting was still intimately tied to the changes in seasonal light and the moon's monthly phases. In Paris, even after the advent of gas lighting in the 1840s, two kinds of lanterns were used, one that burned all night, and another that was lit only when the moon in the streets didn't offer enough light. Even into the early twentieth century, many municipalities planned their lighting schedules in relation to the moon. Long before Paris was the City of Light, it was, as every city in the world, a city of moonlight.

And to enchanting effect. In *Nocturne: A Journey in Search of Moonlight*, James Attlee explains, "Just as in a black-and-white photograph, the lack of colour visible by moonlight makes the architectural structure of the landscape more apparent." Goethe saw this while touring Rome in 1787:

> Nobody who has not taken one can imagine the beauty of a walk through Rome by full moon. All details are swallowed up by the huge masses of light and shadow, and only the biggest and most general outlines are visible. We have just enjoyed three clear and glorious nights. . . . This is the kind of illumination by which to see the Pantheon, the Capitol, the square in front of St. Peter's and many other large squares and streets.

It's an effect we no longer experience today. Too often in modern cities the moon, overwhelmed by artificial lights, makes its crossing unnoticed. And so we forget the wild beauty of this natural light, or worse: We never miss it because we have never had the chance to learn.

Tonight in Austin, Texas, I'm standing on the Congress Avenue Bridge with the chance to learn from Merlin Tuttle, the world's foremost expert on bats. We are here at twilight to see the emergence of between 750,000 and 1.5 million Mexican free-tailed bats from under the bridge, a wild sight in the midst of a city of a similar number of humans. Founder of Bat Conservation International (BCI), Tuttle has worked for more than thirty years around the world to raise human appreciation for these amazing mammals. In the world of wildlife conservation, the name Merlin Tuttle means bats.

In part because we so identify them with darkness, bats have suffered immeasurable persecution at human hands. This persecution has been and is motivated by irrational fears of bats carrying rabies (which only one half of 1 percent do); bats tangling themselves in human hair (which, for an animal whose echolocation can detect a single human hair, is not going to happen "unless you're sporting a head of cucumber beetles," warns one Ohio Department of Natural Resources website); or bats "attacking" humans (Tuttle says in his forty-year career he has yet to see a bat be aggressive toward humans). In the rational world, bats clear our skies of mosquitoes and other pesky insects while pollinating the flowers and fruits we enjoy—tasks that have enormous benefits, economic and otherwise, for humans. Yet around the world—and nowhere more relentlessly than in the United States—colony after colony of bats has been destroyed by humans wielding guns, fireworks, dynamite, flames, napalm, poison, tennis racquets, hockey sticks, and other weaponry, simply because we fear these small, nocturnal flying mammals. In just one example from the early 1960s, the world's largest known bat colony, in southern Arizona, was reduced from thirty million to just thirty thousand, shotgun shells littering the hills outside the colony's cave.

But with his tireless work on their behalf, Tuttle has made a difference. He shares the story of a Tennessee farmer agreeing to let him study the bats in a cave on the farmer's land but telling him, "While you're in there, kill as many of them as you can." When Tuttle found the cave floor littered with the discarded wings of potato beetles—the farmer's main pest—the farmer changed his mind immediately about the bats. Tuttle says that when he first began working with bats and gave presentations, he'd have to spend the first ten minutes of his talk simply debunking myths. Now, he says, he often gives talks during which no one bothers to ask about rabies, or vampires sucking their blood, or the chances of a mass bat attack on an unsuspecting American city.

When Tuttle first moved to Austin in 1986, there probably were some people who thought a mass bat attack was imminent. The Congress Avenue Bridge was only six years old, and a few hundred bats had discovered that its understructure made a perfect roost. Their arrival excited some in the Austin public to call for the bats' eradication. "There was a *USA Today* headline that read, 'Hundreds of Thousands of Rabid Bats Invade and Attack Citizens of Austin,' " Tuttle tells me. "The *Austin American Statesman* had one that read, 'Bat Colony Sinks Teeth into City.' But now that same paper has a bat mascot, and a bat hotline you can call and get information about when the bats are flying." Tuttle laughs, saying that now, when people want to write stories about those days, they can't find anyone who will admit to calling for the bats' eradication.

Tonight as we wait, a crowd of all ages gathers behind the bridge railing. On the Colorado River below, several small tour boats bob with rows of expectant faces, and individual kayakers ready their waterproof camera gear. At one end of the bridge, a billboard advertises Bacardi tequila, prominently featuring the company's bat logo. Close by, bat watchers are greeted by a large black bat statue and a sign proclaiming the Congress Avenue bats to be the "world's largest urban bat colony." And every so often, a "world's largest" waft of "aroma"—the distinct scent of bat guano—comes rising from under the bridge, as though the bats are letting us know they're getting ready.

The tiny percentage of bats that carry rabies wouldn't seem to warrant human hatred toward them. Tuttle tells me that no group of mammals has been more intensively studied for disease issues than bats, and they still turn out to be among the safest animals around. "I rest my case here in Austin," he says. "The public health people said these bats were going to attack people and give them rabies, and the city almost eradicated them. And I said, 'Look, they have value, and if you leave them alone they're going to leave you alone.' And close to thirty years later, we're still waiting for that first bat attack to occur."

One of the ways Tuttle has changed people's perceptions of bats is through photography, a skill he taught himself out of necessity. After writing a feature story for *National Geographic* "talking about what really neat animals bats are and how they're harmless, contrary to people's superstitions," Tuttle says he went to Washington to review photos for the article and found the only photos the magazine had were of snarling bats—a self-defense posture provoked by blowing in a bat's face to get it to open its eyes. The photos made the bats look terrible—"anything but something you'd love." Tuttle told the magazine, "You know, you wouldn't do this to any other kind of animal. People would have a fit if anything was provoked before they were photographed." In the years since, Tuttle's bat photographs have been featured all over the world, photographs that show floppy long or stubby ears, happy eyes, and black, almost see-through wings—in other words, showing that bats are nothing you need hate.

Because there are more than a thousand species of bats—fully one quarter of all the mammal species on the earth—it's hard to generalize about them, but that doesn't keep me from asking Aaron Corcoran to do just that. "Bats are amazing, amazing creatures," says Corcoran, a doctoral student studying bat-moth relationships. When I ask what he thinks most amazing about them, he laughs. "You'd have to listen for an hour. I don't know if there is one thing. What comes to mind first would be just their sensory environment and the world they live in and the speed at which they respond to the environment. An echolocating animal will make anywhere from ten to two hundred echolocation pulses per second, and in the tenth of a second or less between when a sound goes out and when the bat listens to the echoes, it will take all the information, process it, and make a decision about what to do next. The temporal scale they're working at is just incredible. Some of them are making sounds that range from 20 kilohertz to 120 kilohertz, which is six times our range of hearing, in three-thousandths of a second. And then they will hear that sound come back, and just from the properties of that sound tell the texture of an object, how far away it is, and the direction. It can determine very quickly if it's a prey item it wants to chase after, or one it doesn't want."

Donald Griffin, who began working with bats in the late 1930s and is credited with discovering echolocation, said that studying bats was like visiting a magic well. For sixty years, he kept going back to that well, and it never stopped offering new discoveries.

One of the most remarkable stories about bats is the way that, in a cave home to a million or more, a mother can leave to feed and then return to find her own offspring amid pups packed two hundred to five hundred per square foot. According to Tuttle, recent research shows that bats have the same long-lasting social structuring as higher primates and elephants. Researchers have also found that even though the bats they studied went to completely different winter hibernating sites for half the year, they still recognized each other as individuals and had different levels of "friendship." "You can call it whatever you want to call it," Tuttle says, "but it's probably little different than what we have. They're pretty smart."

When the bats begin to emerge into Austin's night in spurts and long flows, a just-audible cheer goes up and people—adults included—actually giggle at the sight. I can tell that Tuttle, having seen the emergence so many times, isn't that impressed. ("It should be getting better," he says.) But I get excited seeing one bat, so the

Hundreds of thousands of Mexican free-tailed bats emerging from under the Congress Avenue Bridge in Austin, Texas. *(© Randy Smith Ltd.)*

ebb and flow of thousands have me whispering amazement. I'm seeing an animal, despite all human attempts to deny it life, come swirling from under a bridge in what seems like joyful flight, and I'm saying to Tuttle, "This might not be good, but it's breathtaking." He laughs. "When it's good, it's one of the wildlife sights of the world. You can see the columns for miles."

The bats swirl in curling black funnels off toward the eastern horizon, headed toward agricultural lands to feed on corn-ear and army-worm moths, which Tuttle tells me are billion-dollar-a-year pests for the Texas economy. In fact, a recent study shows insect-eating bats are worth at least $3 billion to U.S. agriculture alone. As they spend the night feeding, bats eat literally tons of bugs, saving farmers the cost of pesticides. But the $3 billion figure is actually quite conservative, and the savings may be more than $50 billion, as the study omitted many downstream costs to pesticide use, such as human health problems and difficulties related to the development of resistance in bugs. All around the world, bats pollinate fruits and flowers and eat pests that otherwise would devour crops. Ironically, the enormous benefit they bring to human societies is probably proportional to the hatred and fear with which those same societies regard them.

Despite the advocacy of Tuttle and others, bats still desperately need help. In addition to continued human harassment, millions of cave-dwelling bats east of the Mississippi have been wiped out by a plague of white-nose syndrome—a disease named for the white fungus that appears on the muzzles and wings of infected bats—and migrating bats have no defense against wind turbines, which threaten to kill at least sixty thousand a year by 2020 in the United States alone. Unlike birds, which typically die from direct impact with turbine blades, bats are killed by a condition called barotrauma, essentially a version of "the bends" suffered by scuba divers. The rapid drop in air pressure around the wind turbine blades causes bats' lungs to burst. Lights may play a role in these deaths, as bats must fly close to the blades to suffer barotrauma, and some species will be drawn to the insects attracted to the turbine lights. If that weren't enough, European studies show that the glare and trespass of artificial lighting diminishes bat habitat and disrupts their already stressed lives. While scientists struggle to figure out white-nose syndrome and wind turbine deaths, controlling our light would be an easy step toward helping these creatures that do so much to help us.

Tuttle's devotion to helping bats continues to take him around the world. When I'd first reached his house at the start of the evening, he had been memorizing sentences in Spanish for an upcoming talk in Cuba. (He was working on "Newspapers are still, despite all we've learned about bats, publishing stories about how dangerous they are" when I arrived.) At the end of the night, after we leave the bridge, we will drive up the road to look for red bats hunting moths above the lights of the Texas state capitol—the bats devouring the moths in midair, moth wings fluttering to rest at our feet. Tuttle brings his "bat detector" along, a kind of transistor-radio-like device that sputters and beeps and buzzes as it translates bat activities. As we stand listening and looking up into the night, a young woman walks by with a smile, nods at the bat detector, and says, "You guys talking with the bats?"

For Merlin Tuttle, the answer has long been yes, in whatever language it takes.

At twilight on a clear night at the end of June, I park in a Cape Cod National Seashore lot, make my way through hordes of gnats (a woman coming from the beach says to her friend: "Jesus Christ, these gnats are like rabid foxes"), and descend to the sand and the sea. I love the feeling of striding out to meet the night, and this is a night I have been looking forward to ever since reading Henry Beston's *The Outermost House.* Shore fires ahead and behind, birdcalls all around, the pump and splash of crashing surf, the gathering curtains of dusk over the ocean to the east. Finally, I'm walking Beston's beach.

No one I know has written more eloquently about night than Henry Beston. His book, published in 1928, relates the story of his year of living on this beach alone in a small two-room house he designed himself. A frequent summer visitor to the Cape, Beston found in the fall of 1926 that he could not leave, that "the beauty and mystery of this earth and outer sea so possessed and held me that I could not let go." It also didn't hurt that his girlfriend had offered the special motivation of "no book, no marriage." Over the next four seasons Beston paid close attention to "the great rhythms of nature, to-day so dully disregarded"—with special attention to the rhythm of day into night, light into darkness—and called for the recognition of their essential value. "Our fantastic civilization has fallen out of touch with many aspects of nature, and with none more completely than with night," he wrote. "With lights and ever more lights, we drive the holiness and beauty of night back to the forests and the sea."

That Beston was aware of "lights and ever more lights" back in

1928 is remarkably prescient. It would be decades before electric lighting reached much of rural America, and could we step back to see the country then, most of us would not believe our eyes—this was still a very dark land. Yet in addressing his contemporaries, he sounds as if he's addressing us. "To-day's civilization is full of people who have not the slightest notion of the character or the poetry of night, who have never even seen night," he wrote. "Yet to live thus, to know only artificial night, is as absurd and evil as to know only artificial day." Highly attuned to "nature's sustaining and poetic spirit," Beston spent hours walking the beach and reflecting on the natural life he saw there—including a sometimes starry, sometimes moonlit, and always dark sky.

Beston was especially attuned to the birds on Cape Cod. In his second chapter, "Autumn, Ocean, and Birds," he developed his observation of "the lovely sight of the group instantly turned into a constellation of birds, into a fugitive Pleiades whose living stars keep their chance positions," into one of his most memorable passages. He asked, "Are we to believe that these birds, all of them, are *machina*, as Descartes long ago insisted?" Asserting that "we need another and a wiser and perhaps a more mystical concept of animals," he argued that in "a world older and more complete than ours they move finished and complete, gifted with extensions of the senses we have lost or never attained, living by voices we shall never hear." He was writing of the sensory gifts the birds enjoyed long before scientists would make the same argument, his perceptions of the birds simply the result of his paying close attention to his world.

One night after 2:00 a.m., his room "brimming with April moonlight and so still that I could hear the ticking of my watch," Beston walked to the ocean's edge and heard "the lovely, broken, chorusing, bell-like sound—the sound of a great flight of geese going north on a quiet night under the moon." Describing it as "a river of life... flowing that night across the sky," he knew he was witnessing the springtime migration of "the great birds." Here were night-migrating birds using the cover of darkness to make their biannual flights as they had for millennia. He wrote, "There were little flights and great flights, there were times when the sky seemed empty, there were times when it was filled with an immense clamour which died away slowly over ocean. Not unfrequently I heard the sound of wings, and once in a while I could see the birds—they were flying fast—but scarce had I marked them ere they dwindled into a dot of moonlit sky."

Were he alive today, Beston would probably not be surprised to learn that our "fantastic civilization" increasingly disrupts this river of life. In the United States alone, estimates are that at least one hundred million birds die every year as a result of human-made structures. In fact, says Bob Zink, curator of birds for the Bell Museum of Natural History at the University of Minnesota, "estimates range from 100 million to 1 billion birds a year.... Basically, it means we have no idea." ("Most sites are never visited to find dead birds," Travis Longcore told me, "and most of those that are surveyed are visited only sporadically.") What we do know is that some seventy-five million (and growing) communication towers prickle our country's back, most lit and held upright with guy-wires (these wires themselves deadly obstacles), while lighthouses, oil rigs, smokestacks, and wind turbines pepper the land and sea. Most significantly, in urban areas our high-rise buildings create an indecipherable maze for any bird drawn off track. Together these structures present a deadly obstacle course that no bird has evolved to survive, especially at night.

"Nocturnal bird movements are clearly millions of years old," says Andrew Farnsworth, of Cornell University. "The effects of various anthropogenic behaviors, like illuminating what had been, up until a hundred years ago, a dark night sky, can have some potentially dramatic and serious effects." The lights seem to attract and confuse the birds, drawing them toward collisions with human-built structures. Some of the most dramatic instances include a night in 1954 when fifty thousand birds were killed following a beam of light from a Georgia airport straight into the ground; a weekend when more than ten thousand birds collided with smokestacks in Ontario in 1981; another when ten thousand were killed at radio transmission towers in Kansas in 1998; and most recently, in late 2011, a night when more than fifteen hundred grebes migrating over southern Utah apparently were confused by city lights shining off clouds and crashed into parking lots they mistook for ponds. Thankfully, major episodes such as these tend to be the exception. But it is the accumulated carnage of a single bird here, a handful there, a hundred more on a very bad night, that combine to create such a terrible toll. Of the huge number of birds killed, not all can be directly linked to artificial lights, and we are only beginning to understand the exact relationship between artificial lights and bird mortality. But, says Clemson University's Sidney Gauthreaux, "all evidence indicates that the increasing use of artificial light at night is having an adverse effect on populations of birds, particularly those that typically migrate at night."

Within North America alone some four hundred to five hundred different species migrate at night, Farnsworth says. "It spans the taxonomic range, from herons to shorebirds to cuckoos to songbirds. Even some gulls and terns migrate at night, as do a lot of waterfowl and loons and grebes." Many of these species are actually primarily diurnal, and only become nocturnal in their activities during migration season. Farnsworth says that this "season" in fact extends for much of the year, given the many different species of birds involved. "So even though we think about spring and fall migration—and the bulk of it is really April/May, September/October—it's almost all year that there can be movements at night. And that is just in North America. There are definitely quite a number more when you expand that around the world."

But while the sheer numbers of birds killed is dramatic, what's potentially more devastating are specific species of birds killed. In other words, if we're killing five thousand pigeons, that's one thing;

but if it's five thousand of a certain warbler, that's another. "If you're killing a substantial portion of an endangered species at night," Farnsworth tells me, "relative to something like a gray catbird or scarlet tanager or something that's a little bit more abundant or widespread, it's a different scenario." It's for this reason that Farnsworth and others are working to discover the many different strands in the "river of life" flowing overhead. While radar allowed for the discovery of great movements of birds at night, it couldn't tell what kinds of birds were involved. Recent advances in acoustic monitoring technology—using microphones to record the vocalizations of the migrating birds, then using computers to sort the sounds—has allowed Farnsworth and others to begin to reveal the makeup of night migrations. "Each species has a unique vocalization for nocturnal movement," he explains. "Then, you can either listen to them real time, or we have various automated algorithms that can go through it and post-process the data, and that gives you a clue as to what the composition of a nocturnal migration is."

Recently, Farnsworth and a colleague were doing such a monitoring near the Tribute in Light at the September 11 Memorial, on a night when so many birds were drawn to the light that the memorial had to be shut down. "There were thousands of birds up in the beam," he says. "And there was a tremendous amount of calling going on. And as soon as the lights were off, the calling activity dropped to almost zero." The rapid diminishment of the birds' calling was, Farnsworth says, a "striking example of the way various aspects of migratory behaviors change drastically when under the influence of light."

In Toronto, Michael Mesure, founder of the Fatal Light Awareness Program (FLAP), told me that for many years after the city's CN Tower was constructed in 1976, it was lit with spotlights. "On a number of occasions I was onsite observing hundreds, if not thousands, of birds that were circling the structure, trapped in those pencils of light. In fact, the numbers were so great that some were flying right into the concrete, while others were colliding with each other. And then when the lights went off around one in the morning, all those birds that remained trapped in the light just fluttered down to the ground—it was hailing birds everywhere—and it dawned on me, if you're in a brightly lit room and someone flicks a switch, it takes a while for your eyes to adjust. One by one they picked up and flew off into the darkness again."

Within urban areas, perhaps no individual has done more to help night-migrating birds than FLAP's Mesure.

"I had heard about this problem and didn't believe it," he says. "I had to see it for myself. So, I got up bright and early one morning back in 1989, and, lo and behold, I was picking up birds there in the streets of Toronto before day broke." In the years since its founding in 1993, FLAP has made significant progress against bird mortality in Toronto by developing guidelines for architects, engineers, and building owners; having mandatory anticollision measures put into place for all new construction; and helping ensure the development of aesthetically pleasing tinted window-film for corporate structures. Following FLAP's lead, similar groups are active in cities such as New York, Chicago, Minneapolis, and Calgary. Mesure says that while daytime mortality numbers from birds flying into structures are actually higher than night numbers, FLAP has found that the two issues are directly related. "The mass numbers of birds we were picking up after day broke was a result of those birds that were initially drawn into the urban environment at night, and if they managed to evade exhaustion or collision with those lit structures at night, they then had to contend with all the reflective surfaces during the day."

While he uses phrases like "baby steps" and "patience" to talk about the progress made and the work yet to come, Mesure cites two causes for optimism: the rising energy costs of lighting buildings and the new ways of doing business. For example, "the old way of cleaning a structure," he says, "was to come in at the end of the day and sweep through from top to bottom. And the result of doing so was that for about half the night entire structures would be lit." Mesure says that daytime cleaning has become increasingly popular. It once was avoided, due to privacy concerns, but in fact tenants appreciate the chance to establish communication with the people cleaning their offices. As a result, nighttime lighting is no longer necessary.

Mesure argues that the problem of bird mortality in urban areas is not difficult to solve. "If a lake is contaminated or an entire forest is depleted, it will take years of hard work and monies to bring that lake or forest back to life again. But overnight we could solve this problem. How often can you say that?" He says that any person renting space in an office tower has the power to help initiate this change, by requesting changes to building lighting. Already in Toronto, he says, questions about a building's bird-friendliness have become routine for potential tenants. "It's just a matter of time before this becomes how buildings are run," he says.

Mesure's dream is to see all urban centers in Canada and the rest of North America have mandatory measures in place for both new and existing structures to protect migratory birds. He has already seen some of his dream happen, and he doesn't have to go far to be inspired for the work to come. "The moment you start to pick up some of these birds that in many cases people haven't seen in their lifetime, it all comes back. It's a painful reminder of why I'm doing what I'm doing."

Why am I walking Beston's beach? To really see the world he described is to see it after dark. Of course, coming to Cape Cod for a single night (or even a week, a month, or a Beston-like year) won't give me a full understanding of the night Beston knew. And I wonder if what he knew is much diminished, maybe diminished beyond recognition. But that's another reason to visit, to see how much is left.

The beach I walk lies on the other side of Boston's "great yellow sky," opposite of where I was at Walden, so the entire western horizon suffers a sky glow that blots the stars high into the sky. And worse, turning south I see two dozen bright "security" lights on a distant shore. In a community as focused on the natural world as Cape Cod, it's surprising that lazy lighting like this would be allowed. (But this is America, after all. Author David Gessner's compelling essay "Trespassing on Night" tells of a new neighbor on the Cape who insisted upon "plans to illuminate us: thirty-five various spotlights, groundlights, and pool lights," and defended his right to light his property as he pleased by dressing himself in "the scoundrelly cloak of patriotism." Gessner and his wife had been drawn to the Cape by its "feeling of wildness," but, he writes, "here is what all of the new lights did to my wild neighborhood: they tamed it.")

Thankfully, at least on this immediate stretch of national seashore, there are no lights, and I am hoping it won't feel tame. After a while even the shore fires have been doused, and—having made a long detour through Nauset Marsh, past hollow cracked crab shells and the frazzle-feathered bones of long-ago broken birds, around a section of beach closed to protect piping plover and least tern habitat, species Beston would certainly have known—I am nearly alone. Two young fishermen share the beach with me, their poles planted in sand, their lines stretching taut to the surf, where schools of striped bass come close to shore after dark. On the farthest point of land I cross dozens of shorebird footprints in the wet smooth sand, Y-shaped tridents running this way and that.

You can't walk any farther south—the spot of sand where Beston lived lies underwater now, the beach washed away by storms and time. His house, moved back from the edge-of-the-sea setting Beston knew, was named a National Literary Landmark in 1964. With his wife, Elizabeth Coatsworth (her requirements for marriage long satisfied), Beston came to Cape Cod for the last time for the dedication ceremony. He died four years later, and, just years after that, in 1978, an enormously powerful storm swept the house to sea.

The horizon now is just the line of darker presence under a line of dark sky, with silent stars rising from the boom and crash of surf and sea. To the north, the Nauset Light rotates past in regular intervals, to the east a handful of fishing boat lights bob, to the south the lights of insecurity claim the bluffs, and there is the ever-present sky glow to the west. But all in all this is still a dark place, dark enough for a good, deep Milky Way to arch over the ocean, running almost parallel to the beach.

Near the end of his book, near the end of his year, having intimately known "the great rhythms of nature" for nearly four seasons through, Beston wrote of an appreciation for "a sense that the creation is still going on, that the creative forces are as great and as active to-day as they have ever been, and that to-morrow's morning will be as heroic as any of the world."

Though some sand has been washed away, as well as his house and some of the sky, this place at night feels close to what it might have been when Beston was here. I wanted to walk this beach to see if I could sense the power of the old world he'd known, and, though diminished, it is still here—the birds still migrate at night, the schools of fish still move close to shore, the Milky Way bends as it always has. I wanted to see if night here still grows dark enough for such wildness, if it's still possible to step for yourself through a place where the world we will know is still being formed. In the darkness of this sky, the beach, and the sea, it is.

"Learn to reverence night and to put away the vulgar fear of it," wrote Beston. "For, with the banishment of night from the experience of man, there vanishes as well a religious emotion, a poetic mood, which gives depths to the adventure of humanity." When I lie back and close my eyes, this farthest lip of beach right next to the edge of the ocean feels like being up close to an enormous breathing being, the bass drum surf thump reverberating through the sand. Living out here with no lights, alone, you would indeed become sensitive to seasons, rhythms, weather, sounds—right up next to the sea, right up under the sky, like lying close to a lover's skin to hear blood and breath and heartbeat.

4
Know Darkness

But the dark embraces everything:
shapes and shadows, creatures and me,
people, nations—just as they are.
— Rainer Maria Rilke (1903)

Twelve miles of washboard dirt mark the homestretch on the drive to New Mexico's Chaco Culture National Historical Park. My car twirls a dusty white plume as I rumble my way toward the place most refer to simply as Chaco, reaching the park before evening, in time to tour the canyon in what remains of the day's light. Located three and a half hours' drive northwest of Albuquerque, Chaco is famous for the civilization that existed here for three hundred years starting in the mid-800s. Visitors come to wander among the ruins of several "great houses" and kivas that lie spread about the valley, including Pueblo Bonito, the largest of the great houses, built in a half-circle with more than six hundred rooms, many stacked three and four stories high. A circular drive makes most of the main buildings easily accessible, but Chaco's remote location keeps the numbers of visitors low, and it's easy to find yourself standing alone where a bustling city stood ten centuries ago. At Casa Rinconada, the great circular kiva, I watch the sunset ignite ancient stones and canyon walls a burning orange, and with a deep blue sky going ever deeper blue, I hear the dusk's first crickets, and the day's last birdsong, and wonder what it's like out here at night. What must it feel like to be amid the ruins surrounded by the ghosts of those from a thousand years ago? What would it be like to see the corners and courtyards, the small rooms and stones lit only by the moon?

Unfortunately, the canyon closes at sunset, and visitors need to be on the other side of the locked gate soon after. It's reasonable—rangers don't have to worry about visitors becoming lost or hurt in what is even by day an unsupervised canyon, the ruins are protected from damage, and no one has to pay for lights. But for a park that celebrates a culture that kept close watch on the night sky, the fact that the canyon is off-limits at night feels awkward, regrettable, a missed opportunity. At least, that's my initial impression.

I first came to Chaco fifteen years ago, just after I had moved to New Mexico from Minnesota, and I was still making sense of the desert—the mesas and mountains, the canyons river-washed and rusty-hued, bluebird blue sky mornings and, of course, green chile. I came to love all those things, and I still do. But while I certainly remember the bumpy last few miles to Chaco, I realize I have forgotten the unique quality desert ruins hold, the sense that with the rock, air, and light so much as it must have been, it's as though the former residents only recently left. I'm happy to feel that again, because I've come this time wanting to place myself where a culture knew darkness more intimately than we do today.

One of the biggest draws for visitors to Chaco is that most of the structures seem to have been built with astronomical and lunar alignment in mind. Many of the pictographs found on canyon and building walls seem to represent astronomical events, including one known as the "supernova pictograph" thought to depict an event from 1054. But in fact, no one really knows. And so, while the structures and alignments draw visitors, the real attraction here is mystery.

"It probably would be better labeled the 'maybe supernova pictograph,' because there's no way to be certain," says Angie Richman. An archaeoastronomer with the National Park Service—which means she is particularly interested in the cultural role the sky played in ancient cultures like Chaco's—Richman says that, nonetheless, "there's enough around Chaco that shows that they were watching the sky enough to notice changes. There's rock art that's potentially comets and eclipses, and the sky clearly had significance for them in their day-to-day life, for keeping track of time and knowing when to plant and harvest crops. And it was also spiritual—looking at the night sky and seeing the sun and the moon as gods and the stars as something to help guide them was infused into every aspect of who they were."

The ruins of Casa Rinconada stand across the canyon from Pueblo Bonito and just a little farther east, one of the largest kivas in the Southwest, and a prime example of the kind of solar alignment that so interests visitors. Twenty-eight small square niches line the inside walls of the circular building, and two large T-shaped openings for doors lie opposite each other on an exact north-south line. Each summer solstice, if the sun is shining it sends a beam of light through an opening in the eastern wall directly into one of the twenty-eight niches. This apparent solstice alignment draws visitors from all over the world. No one really knows whether this occurrence is actually intentional or just a coincidence; the kiva walls were heavily reconstructed in the 1930s (and by Indiana Jones lookalikes in baggy pants and pith helmets, no less). Still, while some of the architectural feats seem remarkable for a culture that flourished so long ago, if your whole existence rested on your ability to understand the movement of the stars and moon, the sun and the seasons, you would probably figure out a lot of things that would seem mysterious to modern observers for whom those movements have become almost irrelevant.

While Chaco doesn't compare in terms of height or curves to

A slow shutter captures Chaco Canyon's Casa Rinconada at night. *(Tyler Nordgren)*

many other canyons in the Southwest, for a place to watch the sky, it's hard to imagine a better setting. The canyon runs east to west, narrowing at either end, widening in the middle, flanked by flat-faced sandstone cliffs of matching height. And so, for much of the year, the sun and the moon both rise at one end of the canyon and set at the other. The canyon itself is wide enough that, while the walls create the horizon, you have a broader view of the sky than you would in a narrower, deeper canyon. The effect: In Chaco Canyon you stand as though in a stadium built to watch the sky, or an enormous ancient planetarium. It's easy to imagine Chacoans lying on their backs watching the three-dimensional universe above, with stars coming down to Earth, or shooting from one end of the canyon to the other.

The Historical Park has long focused on protecting the building ruins, but an increasingly important mission is to protect the Chaco culture's connection to the sky. Along with G. B. Cornucopia, a ranger at Chaco for more than twenty-five years, Richman has been instrumental in making this piece of Chaco's ancient heritage part of its present-day appeal. Beginning in 1998, she and Cornucopia began to offer nightly astronomy programs near the park's visitors center—just outside the locked gate to the road into the ruins. Thanks to contributed telescopes and the volunteer efforts of amateur astronomers, the park now offers several different nighttime programs each week.

The night of my visit, Cornucopia entertains a crowd of several dozen visitors with time-lapse images of the moon's cycle like a film of a beating heart, and a view through the park's largest telescope of the M13 globular cluster—a tight grouping of several hundred thousand stars—that looks like a sparkling snowball, light as air. The starry sky rises right from the black silhouette of the canyon wall behind us. It's a sky Cornucopia calls "our most direct link" with Chaco's ancient culture, "the same sky it was a thousand years ago." And despite increasing light pollution from surrounding communities, Chaco is, he says, "still a very dark place," a place that, out among the ruins at night, can be eerie, and can make you feel like "you don't know what century you're in."

I am certain I'm not alone among the visitors wishing I could venture out to see it. But the longer I'm here, the more it feels right to leave the ruins alone. Especially if Cornucopia is right, that it's at night that Chaco is closest to what it was, it actually feels better to stay behind the gate—out of respect for those who once lived here, yes, but also because leaving what it feels like in Pueblo Bonito or Casa Rinconada under a big moon or star-plush sky a mystery actually adds to Chaco's appeal. At Walden, it felt right and respectful to visit Thoreau's cabin site at night, as though in doing so I connected with some part of myself. Here, while I'm pleased to visit, there's a sense this place isn't mine, and so to leave it unknown and unvisited feels anything but awkward, regrettable, or a missed opportunity. As the crowd begins to thin and the telescopes are put away, I look west into the canyon and imagine again the ancient Chacoans gazing at the unfathomable sky. The missed opportunity would be if we were to always get what we wanted, if we weren't to leave some of the night alone.

In his elegiac *In Praise of Shadows*, the Japanese novelist Jun'ichirō Tanizaki wrote of the Westerner, "from candle to oil lamp, oil lamp to gaslight, gaslight to electric light—his quest for a brighter light never ceases, he spares no pains to eradicate even the minutest shadow." Written in 1933, as the electric light washing over Japan threatened what Tanizaki saw as the vital role of darkness in Japanese culture, *In Praise of Shadows* reads as though it could have been written yesterday. While he does not object to the modern conveniences of electric lights or heating or toilets ("truly a place of spiritual repose"), Tanizaki does want us to recognize the "senseless and extravagant use of lights" that has "destroyed the beauty" and "this world of shadows we are losing." I'm struck by Tanizaki's critique of the Western mind-set, and by his wondering whether, if the Orient had developed its own science, "the facts we are now taught concerning the nature and function of light...might well have presented themselves in different form."

That a culture might think of darkness differently than we in the West do today stays with me. But we don't have to look far—either back in time or across the ocean—to find different cultural attitudes toward darkness and night.

While it's impossible not to generalize when speaking about the philosophy of more than five hundred different nations, broadly speaking, the way night is represented in Western culture is very different from the way it's represented in the varied cultures of Native North America. While in the West we've been locking our doors and shutting our windows against the terrors of the night, both natural and supernatural (think werewolves and vampires), for centuries, Native American cultures, says Joseph Bruchac, have been seeing great spirituality in the darkness. "When we go into the sweat lodges, for example, we go into darkness. We go back into the womb of our mother—that sense of darkness being embracing and protecting. And when we look at the night sky, the Milky Way, the road of stars is the passage of souls. It is the way from life to afterlife." An Abenaki storyteller and author of more than seventy books, Bruchac says that in traditional Native culture night is often seen as a time of healing, that many ceremonies and rituals take place at night, and that there is a sense of possibility in the night sky. "So when we look at the night sky we see many, many things," he says, laughing. "We're not looking for some shape darkening the moon and coming to suck our blood."

While Western culture heavily emphasizes good versus evil, "American Indian cultures tend to be much more ambiguous," says Bruchac, "or at least more broad in their view of the role of different creatures and different things. There are gradations of gray, and you wouldn't really hardly ever say anything was just plain evil. It might not be good, but not evil in that European sense of irrevocably beyond the pale." Even the concept of darkness and light as separate is often quite different, too. Black is not always bad, white not always good. "The two balance each other. It's almost a yin-yang kind of thing," Bruchac explains. "Certainly in Abenaki traditions, the hero Gluskabe is often represented as having a white wolf by one side, a black wolf by the other. One is day, one is night, and they're equally important guardians or companions, for him and for all people."

Of course, Bruchac knows that a strong traditional relationship with night doesn't automatically translate to the modern world. In fact, in many cases, he says, that connection to night is being lost. "It's less common," he tells me. "You've got reservations out West where gang activity is more common than going out at night with your family." Visit eastern Arizona's Canyon de Chelly, a national monument famous for ancient cliff dwellings built into the long sway of a canyon wall, where a modern Navajo community lives "connected to a landscape of great historical and spiritual significance" (as the monument brochure says), and you may be stunned to see the same unshielded security lights that blare from barnyards and driveways across America. The fact that the Navajo here are "connected" to this landscape of "spiritual significance" apparently does not make them any more sensitive to light pollution than the rest of us.

And yet for many in the Native community, the tradition remains quite alive. As the Iroquois writer Doug George told me, "We have the night so the earth can rest. We have night so we can see the star path called the Milky Way and trace our beginnings to the Seven Dancers, the Pleiades. We have the night so some spirits may wander about and sense the physical life from which they came. We have the night so we can release our spirits to travel across space and time, to visit with other spirits and gain their advice. It is by night that we cross into other worlds, other times past and future. Our physical bodies need to dream, as this reality is only one of an infinity of realities, and only through dreaming are we able to make peace with this fact. We are never alone; nor are we restricted by the body as long as we use the night to see our place in the right perspective."

Part of that perspective, says Bruchac, is a spiritual understanding of darkness completing the light within Native cultural traditions. "You need to be able to fall into darkness," he explains, "but you need to be able to climb out of it at the other end, to recognize that it is a cycle. Just like day becomes night, and night becomes day. The cycle continues. We need to go through that as people, too. If we try to make things too easy for ourselves, for our children, and for our culture, we are making a great mistake, because as a Cheyenne elder once said to me, 'Life is supposed to be hard. Otherwise, we wouldn't appreciate it at all.' "

* * *

Eric G. Wilson is about as successful as an American college professor can be. Only in his midforties, he is already a tenured faculty member with an endowed chair and several books to his credit. He is also, without hesitation, against happiness. At least, that is what the title of his book would have you believe. In *Against Happiness: In Praise of Melancholy*, Wilson makes a stand against the kind of addiction to happiness so prevalent in American society. He wonders if "the wide array of antidepressants will one day make sweet sorrow a thing of the past…if we will become a society of self-satisfied smiles. Treacly expressions will be painted on our faces as we parade through the pastel aisles. Bedazzling neon will spotlight our way." With artificial light, Wilson writes, "We are right at this moment annihilating melancholia."

What do we mean by the word "melancholy"? ("Melancholia" is its more traditional version.) A modern dictionary offers synonyms such as "sad," "gloomy," and even "depressed," all states that, these days, doctors commonly treat with pills—Prozac nation and all that. But it wasn't always so. Look at versions, from the 1950s and 1960s, of the *DSM*, the manual psychologists and psychiatrists use to diagnose mental health, and you will see a place for the natural human emotions of mourning, sorrow, grieving—all states of a natural melancholy. Too often now, critics like Wilson claim, we treat these natural states as clinical depression, something to be medicated away. "Either you're happy or you're clinically depressed," Wilson tells me. "There's no place for that middle ground which is obviously highly significant. For me melancholy is inevitable. I wanted to reclaim that ground against happiness."

A professor of Romantic literature, Wilson weaves his thoughts with references to the eighteenth- and nineteenth-century poets he's studied for years—Blake, Wordsworth, Dickinson, Keats. When I contacted him for this book, I thought what better place to meet than in his office, surrounded by the works he loves. After all, you might think the setting would be appropriate for a scholar of melancholy—that I'd knock and push open the door to a dark chamber lit only with candles, incense in the air, an organ fugue from dusty speakers, a battered desk awash with loose papers and ancient verse, the gloomy professor hunched over tortured words. But Wilson suggested we meet at a trendy bar in the city's art district, out in the open, surrounded by young professionals enjoying the evening. It turns out the man who wrote "to desire only happiness in a world undoubtedly tragic is to become inauthentic, to settle for unrealistic abstractions that ignore concrete situations," also enjoys good local beer.

In fact, to meet in public surrounded by people having a good time fits perfectly with the argument that melancholy is, as Wilson says, "an essential part of a full life"; that, rather than see melancholy as some sort of failure or unfortunate disease, we can see this dark quality as being as natural to a human life as the dusk sifting down outside the bar windows. Similarly, one way to see the Romantic emphasis on the value of melancholy is as a response to the excesses of the Enlightenment. "In terms of literary history, the late eighteenth century is the time when poets started saying, look, to emphasize only reason is to miss deeply rich experiences that connect us to the meaning of life," Wilson explains. "Blake was horrified by Newton, because Newton thought that you could reduce the world to atoms moving through the void with mathematical predictability. Basically, the world was a machine. Obviously Romanticism is a very diverse literary concept, and I'm not trying to say it's only one thing. But I would say most of the major figures were keen on feeling, emotion, melancholy, darkness, chaos, possibility, freedom—all ideas connected to twilight or night."

Wilson describes melancholy as "an active longing for a richer relationship to the world than we have enjoyed before." He draws on the English poet John Keats's 1819 "Ode on Melancholy" as an example. "Keats says the only way you can really appreciate the world in all of its complexity and beauty is to feel sorrow at the fact that everything is passing. So if I hold up a porcelain rose it's not as beautiful as a real rose. Why is a real rose beautiful? Because it's transient, it's fragile, it's tender, it's decaying right before our eyes." For Keats, the aesthetic appreciation of the world comes out of a deep sense that everything is, as Wilson says, "passing into darkness," and therefore we long for things to stay, not to die. But in that very longing for things to stay comes a desire to embrace them more intensely. For the Romantic poet, Wilson argues, "melancholia over time's passing is the proper stance for beholding beauty."

When Wilson calls melancholy "a twilight state" between the artificial light of our obsession with happiness and the "deep, obliterating" darkness of depression, he knows of what he speaks. To go with his tremendous professional success and his roles as husband and father, Wilson has battled depression most of his life. In his memoir *The Mercy of Eternity* he describes a "despair so deep" that he "was worse than dead. I was neither dead nor alive, neither restful nor energetic. I hovered somewhere in between, a ghost." For me, knowing this about him makes his reflections on melancholy all the more meaningful. For, while it's easy to repeat clichés such as "Whatever doesn't kill you makes you stronger," the truth behind these phrases has been with us forever: Our most difficult experiences carve in us our deepest understandings of life. "I feel as though I were authentic, true, alive," Wilson writes of melancholy. "All fakeness falls away, and I am at the core of life."

The value of metaphorical darkness is told everywhere if we look for it—in our poetry, religion, literature, art. But the key word here is "if." Everyone experiences the darkness of difficult times—if not depression, then loss of infinite kinds, including simply the everyday passing of time. To think that melancholy—which seems a natural response to the coexisting realities of beauty and mortality—is the same as clinical depression is tragically mistaken. Words like "sad," "gloomy," and "depressed" leave no room for the rich, dark quality of melancholy, which I've always seen as a sensitive appreciation that change is happening every second of our lives, that everything and everyone we love will die, and that in knowing this we have the opportunity to share our gratitude while we still do have time.

"I think that when we're truly moved by something, it always feels sad," Wilson says. "And it may not even be sad. . . . I love this folk band the Carolina Chocolate Drops, that old-time string band music. I saw them in Greensboro two weeks ago, and during some of their songs, I felt myself tearing up. It was really this sense of, life is fucking large and marvelous and weird and I don't even come close to getting it. And I love that. I feel like something deep and inscrutable opens up for us when we see something beautiful. And there is that sense that, yes, this is transient. It will never be again. But there's something else going on, too. It's a darkening, but a darkening that suggests there's more. It's like the terra incognita, the unknown land on the map. I think that's what the darkness is: We have places within us which can never be mapped."

I have been to the South Rim of the Grand Canyon in the past, seen the view, seen the smog, thought, *Shoot, this isn't right*, and gone on my way. But I have not been to the North Rim before, and I have not visited this grand old park in search of the night.

I have plans to meet the full moon at its rise, and I will not be late. Coming into the park from the north through the beautiful meadows of the Kaibab Plateau, I pass herds of bison and stands of ponderosa pine, find a campsite, take in the canyon with the late afternoon crowds, enjoy dinner at the lodge before sunset, and saunter the paved path toward an east-facing lookout. I step from the path to climb gnarled beige rocks, and my lookout takes some scrambling to reach, but it's not that far from the lodge, with its amber-lit windows against the fading blue twilight. The curve of night closes the day sky to a small half-dome in the west, and my fleece jacket feels good. The few lights from the South Rim seem the same size as stars, though they are pink from high-pressure sodium. They and a jet's blink and contrail are the only signs of the human.

The moon first appears as a fire-orange flare behind the flat horizon mesa, then, like a red-pink ball burning toward me, devouring forest and moving west. Technically, we are revolving toward night at something close to 1,000 mph, but we never notice that. What we notice, if we notice, is the speed at which the moon rises: slow enough to make you impatient if you're still on human time, but still fast enough you can see it happening. With the entire glowing ball well above the horizon, lighting these gnarled rocks beige-white, the moon seems smaller here than elsewhere, and the sky enormous. Then, I know why: I can see almost entirely around me, only back toward the lodge do rocks and pines disrupt the 360-degree flat horizon. Otherwise, my perch is a ship's tower in the middle of the sea, smooth horizon all around, and above, a full bowl of stars. The view is vertigo-inducing and makes me wobble-kneed, swervy. I lie back on the rocks, rocks that in daylight appear full of ocean creature fossils, lie on an ocean floor looking up at the night sky.

During the day, there were obese Americans complaining about having to walk a hundred feet, a young French couple with their toddler in a backpack, a British girl clutching a golden teddy bear and telling it to not look down. But at night, now, almost no one—only two couples—shares my view of this grand moon rising over the full canyon, and I know of their presence only by their camera's occasional click and flash. At night now in natural light the layers of rock are clearer, the sense of eternal time greater—the ancient moon on the ancient stone, and all of us just passing through.

Here, too, is the desert quiet—and quiet is a quality closely related to darkness. At least it ought to be. The relationship between light pollution and noise is such that if you find one, you likely find the other.

For me, the quiet of night has long been a friend. In college I would listen to my radio after turning out the light, my ear to the speaker so as not to wake my roommate. On Minnesota Public Radio from 11:00 p.m. to 5:00 a.m., Arthur Hain hosted *Music Through the Night*, his low, calm voice perfect for nighttime radio. Lying in bed listening, I traveled to nights across oceans of water and time, once again a boy in the basement bedroom of my grandparents' southern Illinois home, or eighteen and walking thousand-year-old European streets, or back up north at the lake, standing on the dock beneath great swaths of stars. Beneath the sun, the magic faded. During the day my little radio's music sounded canned, artificial. But at night the very quality of the tiny speaker's condensed sounds brought them close, as though I listened through floorboards to secrets whispered at a dinner party below.

The three tiniest bones in the human body—the hammer, the anvil, and the stirrup—deliver auditory miracles all day long. So often we take them for granted and hardly notice. At night, I have learned to notice. Through my little radio, through my time outside, I have learned that the natural sounds of night are solitary, singled out, floating. Sometimes they seem meant only for me.

I am reminded of James Galvin's beautiful memoir *The Meadow*. Galvin writes that his neighbor Lyle "told me he could hear different tones emitted by different stars on the stillest, coldest winter nights. He said he could tell which notes came from which stars. He couldn't hear them all the time, just winter nights, and then, when he was about sixty, he admitted sadly that he couldn't hear them anymore. Age, I guess."

Sometimes, on a desert night like this, or on clear winter lake nights when stars stand in three-dimensional beauty—closer stars closer, farther stars farther away, and it feels as though I could reach my arms into them, or that falling from Earth would mean falling among them—I see how someone could hear stars. And I wonder, do I miss hearing stars because I don't get myself into dark country often enough, or live surrounded by noise, or simply because I don't pay attention?

Our world is filled with noise, which not only robs us of beauty but as an environmental cause of ill health is second only to air pollution. Exposure to excessive noise has been shown to raise blood pressure, disturb sleep, and stress us into illness. In response, the European Commission—the executive body of the European Union—has set guidelines for maximum levels of nighttime noise at 40 decibels, about as loud as a library. In the United States we lag far behind in protecting the quiet of our nights. President Reagan abolished the Environmental Protection Agency's noise program in 1982, and since then there has been little action from the federal government to protect our country's quiet nights.

Twenty-four hours a day in our cities the gathered growl of countless engines surrounds us, and even in forests and mountains and on country lanes, the solitary engine of a single vehicle or overhead plane interrupts the otherwise natural quiet. At least with noise—we haven't yet reached this point with light—if your neighbors get out of hand, the police will respond to your call.

At least with a night like this, in a place like this, protected for every citizen in the world, there is still quiet.

I first met David Saetre when I moved to the northern Wisconsin town of Ashland to teach at the small college where he is campus minister and professor of religion. But those titles only hint at the vital role he plays in the community, both on campus and in town. Last spring, for example, when in the course of barely a month the gregarious dean of the college died of cancer and a beloved student, a senior I knew well, was struck and killed by a car on frozen Chequamegon Bay after midnight, the community leaned on

Saetre in its grief. David Saetre is the kind of person who lives in every community across the country, across the world: a no one to outsiders, and everything to those inside. I know him as a deeply joyful, conscientious man, and I knew I wanted to talk with him especially about metaphorical darkness and light.

"I grew up on the edge of a small town, and there was a kind of freedom to rural small community childhood that I think we have lost, by and large," he tells me. "I can remember from a very early age playing in the dark and being allowed to roam until bedtime in the dark. It seems to me that very few parents would allow children to do that kind of thing today. Not only was there a freedom to that, but there was a kind of acquaintance. I tell the same story if someone asks me about care and concern for the earth. Real care is a form of intimacy, and you have to cultivate that intimacy somehow. I was lucky to have grown up with it literally—in the woods, in the soil, in the earth. The same is true with the dark."

Listening to Saetre talk about his childhood, I think of Richard Louv's book *Last Child in the Woods* and his argument that American children are now living a "de-natured childhood" resulting in a "nature-deficit disorder" that has serious consequences for both the children's health and our society as a whole. We could say the same thing about children and darkness. As Joseph Bruchac told me, "I think we could talk about a darkness deficit, yes." If we never have the chance to know literal darkness as a child—to play in that darkness—it would make sense that we would grow into adulthood without appreciation for either the literal darkness of our nights or the figurative darkness of our lives. As Saetre says, "We are not taught that not knowing is okay."

Despite his official roles, Saetre has a love-hate relationship with organized religion. "My problem with Christianity and with most of the structured religions," he explains, "is the attempt to say too much. They claim too much, especially in terms of what is clear, in attempting to destroy essential ambiguity." He calls this an "obsession with a kind of false clarity," the idea that everything can be "brought into some kind of pure light." He continues to be drawn to the study of religion, he says, as a way to "try to understand those encounters with the paradoxical ambiguities of our lives that I think really are at the heart of our human experience." In fact, he sees his role in the community as not only to suggest the possibility of the sacred in people's lives but also to "maintain the dimension of ambiguity or of the question—the essential character of doubt."

Doubt. Not knowing for sure. Being open to ambiguities. "Loving the question," as Rilke said. Saetre had better not run for political office, at least not in the United States. We are not very good at remaining open to doubt in this culture, not very good at not knowing. We are very good, on the other hand, at the bumper sticker, that sticky product that reduces a complex issue to something simplistic, black and white, right and wrong. We want answers, the shorter the better.

Here's a bumper sticker almost anyone would understand: "Light good, dark bad." It's the classic understanding of Christian theology, one that reflects the false clarity that concerns Saetre. It also happens to be woefully incomplete and overly simplistic.

In truth, the powerful metaphor of light versus dark—of dark as sin, light as good, dark as evil, light as godly—comes out of only a part of the Judeo-Christian tradition. If you think about the experience of light and dark in the stories of the Bible, a different picture starts to emerge. In the Old Testament, for example, the night—darkness—is frequently the place where people experience the presence of God.

Think of Genesis 32, the story of Jacob wrestling all night with "a stranger" or "an angel." Right before the break of day he gets the stranger in some kind of hold and says, "Bless me before I let you go." The stranger blesses him and gives him a new name—in other words, a new identity. Jacob becomes Israel in that story. The way the story is usually interpreted is that Jacob wrestles with God, signifying that night is the time when humans encounter God in God's most existentially vivid life-changing form.

Or think of the first book of Samuel, chapter three. Samuel is a little boy, and he wakes in the night to a voice saying, "Samuel, where are you?" He runs to his father, thinking that he is calling him. This happens three times. Finally, his father tells him, It's not me who is calling you; the next time you hear the voice, be still. It happens again, and of course it's God, calling Samuel to be a prophet.

"In so much biblical narrative," Saetre says, "night and the experience of dark is not the place of evil and sin but rather the place where humans encounter the deep mystery of being. There's something about the deprivation of light that allows the characters in these narratives to experience reality in its most profound and holy form."

This is such a different picture of the Judeo-Christian tradition. There are many other examples: Jesus's most profound experience of God as his Father occurs in the garden of Gethsemane at night. Jesus is referred to in the Gospels of Matthew, Mark, and Luke again and again and again as going out into the desert to pray at night. Even in the greatest of the stories of the early Hebrew people—the story of Passover, and the Exodus—the angel of death comes when? At night.

"From a Jungian archetypal analysis," Saetre tells me, "what the story of the Exodus is really about is a dying to the old ways, which are slavery, and from which a whole people is liberated." I like his phrase, "a dying to the old ways." He is arguing that night offers us the opportunity to leave old ways that have us enslaved—an opportunity to change our lives. As he says, "night is the time of liberation, the time and the place where we are set free from the overbearing presence of light. Or, in other words, sometimes light keeps us from experiencing the deep truth of things."

Maybe the best example from Christianity of night's being a time of significant experience comes in the literature of the sixteenth-century Spanish monk St. John of the Cross, from whom we have the phrase "dark night of the soul." His first poem, "On a dark night," is his most famous. St. John described this sensual poem as the product of divine inspiration.

Saetre loves St. John's work. "In his poem, St. John wrote that his experience took place 'on a dark night,' with 'my house, at last, grown still.' The house of our lives is mostly experienced in daytime, and the daytime here is the life of obligation, and the life of obligation is so overbearing that it becomes oppressive. In order to experience the liberation of transformation—being loved this deeply transforms him, or, as St. John says, 'Lover transformed in Beloved!'—we need the dark of night, because the daytime is so filled with the burdens of responsibility. The light throws us back into all the stereotype false selves that we put on, all the masks that we wear in order to fulfill obligations."

Saetre smiles. "And we all have to do this—you can't live all the time in the night. But our daylight selves are not our full selves."

The dominant metaphor in the West is still that light is good and dark is bad. But Saetre argues that the night (and its darkness) is the place where the soul encounters the true self and transforming love, and the day (and its light) is the place of, as he says, a false clarity filled with burdens and toil and responsibilities. This understanding doesn't associate light with evil—it's not that old dualism at all—but instead argues that we need both light and dark, and that because daylight is the place of obligation, to experience the true self requires the night.

But what about God declaring, "Let there be light"? Saetre tells me that this famous passage from Genesis isn't as clear as people think. In fact, "Genesis suggests that darkness precedes light, and it is only out of the dark that the creative impulse of that which we call

God emerges, as if to say that darkness then contains an essential and necessary element in the creative process." But does that mean Genesis gives us license to light up the world? "That is a false step," he says. "It's not inferred from the scripture."

Somehow I have a feeling Saetre's message hasn't reached Spain. While in the sixteenth century the cities in which St. John wrote and lived would have been as dark as any on earth, today they are as bright, lit with an overabundance of electric light. First in Toledo, outside Madrid, where John was imprisoned—the root of his own "dark night" experience—and next in the southern city of Granada, where he wrote his poem in the Alhambra's shadow, I walked where St. John walked, wishing I could experience the inspiration for the "dark night of the soul." And, even though I expected these Spanish cities to be lit as brightly as they were, it was still disappointing. Toledo, especially—the stone city on a hill; a UNESCO World Heritage Site, where the bells of its immense neighborhood cathedral echo along narrow winding streets—could be such a lovely nighttime attraction if its lights were under control. Instead, they glare as though lighting any city anywhere. For a poet writing today, there would be no great experience of literal darkness for inspiration as there was for St. John centuries ago. Somehow a milky-gray-washed-out-night-of-the-soul doesn't have the same ring. And for future readers of St. John's work, I wonder if eventually the idea of a "dark night of the soul" will even make sense.

Wanting to know more, I ask Saetre to return to one of my first questions. "How do I see my role as minister? I talked about making room for the sacred in the ordinary, and I said creating space for doubt, because I think doubt is essential to religious life rather than antithetical—there is no faith without doubt. The opposite of faith is not doubt; the opposite of faith is certainty. But the third thing I would add is that I am sought out—sometimes it's the community at critical moments, sometimes it's individuals in their solitude—to try to find ways to help us know our sadness, know our sorrows and our losses, and incorporate them into our ongoing journey, into our lives. How do we avoid these twins of denial, on the one hand, or triumphalism, on the other, and allow ourselves to be vulnerable in this moment of sadness and sorrow? I think that's what ministry is in our time and maybe in most times."

As I listen to Saetre, I know he's talking about more immediate sadness, such as the deaths of our student and our dean, but I'm thinking, too, of the ecologist Aldo Leopold, and a particular passage from his 1949 book *A Sand County Almanac* in which he mourned the destruction of the natural world. "One of the penalties of an ecological education," wrote Leopold,

> is that one lives alone in a world of wounds. Much of the damage inflicted on land is quite invisible to laymen. An ecologist must either harden his shell and make believe the consequences of science are none of his business, or he must be the doctor who sees the marks of death in a community that believes itself well and does not want to be told otherwise.

Leopold wrote *A Sand County Almanac* in the last years of a life cut short—he was only sixty when he died—and his book is a record of his experiences and philosophy. We think of him now as the "father of ecology" and the "father of wildlife management," and he continues to influence conservation thinking more than sixty years after his death. In his comprehensive study *Wilderness and the American Mind*, Roderick Nash titled his chapter on Leopold simply "Prophet." Leopold saw what others did not see or did not want to see. And with that vision came a fair amount of sadness. For how can you love something and not mourn its destruction?

Interestingly, when Leopold first drafted *A Sand County Almanac*, he placed the above passage in his introduction, but then he moved it, fearing its somber message might turn people off. But at least for me—and I'm sure for Leopold—the answer has never been to "make believe" that the realities of human destruction don't exist. Our oceans are on the brink of collapse, our land is full of poisons, our world is warming at a terrible rate—how do we live a joyful life while still engaged with this knowledge?

Most days I live awed by the world we have still, rather than mourning the worlds we have lost. The bandit mask of a cedar waxwing on a bare branch a few feet away; the clear, bright sun of a frozen winter noon; the rise of Orion in the eastern evening sky—every day, every night, I give thanks for another chance to notice. I see beauty everywhere, so much beauty I often speak it aloud. So much beauty I often laugh, and my day is made.

Still, if you wanted to, I think, you could feel sadness without end. I'm not even talking about hungry children or domestic violence or endless wars between supposedly grown men, though these are certainly worthy. I'm just thinking about what's happening to the natural world. But "you mustn't be frightened if a sadness rises in front of you, larger than any you have ever seen," said Rilke. "You must realize that something is happening to you, that life has not forgotten you, that it holds you in its hand and will not let you fall."

I tell Saetre of Leopold, of Rilke, of how, already in Australia, they're speaking of *solastalgia*, about missing a loved place that still exists but to which the old birds and plants and animals no longer come. A word newly coined for our time, *solastalgia* combines the Latin word for comfort (*solacium*) and the Greek root meaning pain (*algia*) and differs from nostalgia in that it's a yearning for a place you still inhabit rather than one you've left behind. It's a word we'll be hearing more often, for wherever we live, the climate has changed, or soon will. Next to my own death or that of my family this is the darkness I fear most, this sadness at the ongoing destruction of the wild world.

"It's okay," he says. "You can survive being that vulnerable. You really can. I think it is at the heart of being human, sorrow and sadness. It's certainly an essential component of loving. If you don't want to cry, then don't love anything."

One night David Saetre was ten years old and walking home at night in the dark, and he walked into the Lutheran church where his family worshiped. The church was unlocked, dark; there was no one there. He walked right up to the altar. "I was still pretty young," he says, "because all I knew was that the altars in church were 'the holy.' You didn't go there. As far as I knew I might be struck down dead, I might be violating something sacred. I remember feeling the paradox of exhilaration and fear. Much, much later, the first time I read Rudolf Otto's *The Idea of the Holy*, in which he describes religious experience as the *mysterium tremendum et fascinans*—that is, the encounter with the mystery that causes you to tremble with fear and yet is so fascinating or compelling that you cannot help being called to it—I said, 'I know what that is.' To stand in the presence and tremble and remain fascinated without being consumed, that is just the classic religious experience of holy fear."

"Does that relate to our fear of the dark?" I ask.

"It is precisely pertinent to how we regard darkness," he explains, "because it has represented, especially in the West, the ultimate encounter with the dark and our need to again find an unambiguous light so that we don't have to face the kind of holy fear of death. What I would call a necessary fear."

"Could you say 'good' fear?"

"Yes, good fear, valuable fear. It makes me think of the No Fear brand of clothing that was so popular awhile back. I asked a student one day, 'What does that mean to you? Why are you wearing that logo?' And some of the students then started talking about taking risks, and to be really alive you have to have no fear. And I said, 'Bullshit. If you are really, really alive, you are scared shitless and you do it anyway. You get on that surfboard, you climb that wall. If you have no fear, you have no experience. So absolutely take the risks: Go backpacking, go whitewater kayaking, and know the fear. In fact, put that on your shirt: Know Fear.' "

"It's the same thing with sadness," I say.

"Instead of no sadness it is 'know sadness.' If you don't become intimate with sadness, you cannot be intimate with yourself or others or the world."

"Know darkness?"

"Absolutely."

In this northern Wisconsin town, sometimes driving for morning coffee I would pass a red fox reduced to slop at the road's edge, or a broken fawn, its stick legs and crumpled white-spotted coat reminding me of bagpipes dropped in a corner. One spring a lumber truck demolished a bear, leaving dark blood and black fur smeared across several yellow highway dashes.

Death here is not always so dramatic—creatures as tiny as mosquitoes and dragonflies, as small as toads and rabbits and turtles, turn up regularly. Fish float to shore, their mouths fixed in last gasp. A wolverine's forearm, lying by a two-track through the woods, claw still attached.

But when it comes to human death, my culture hides it as much as possible, treating it similarly to melancholy or sadness, or darkness—a subject to be avoided, rather than a part of human life as natural as the moon and the tides.

I have never been with another human as he or she died. My grandparents all died far from me, with my seeing them alive for the last time and seeing them again at their funeral separated by a moat of unspoken space. I do not complain; I am grateful that death has been an infrequent visitor. I know it's a matter of time. But when I receive an email from a dear colleague across the country telling me she has been diagnosed with a rare and aggressive form of cancer, that she will begin chemotherapy this fall, I bow my head and think, *What do I know about death?*

"Well, don't be too hard on yourself," Saetre says. "The task isn't to overcome the anxiety. It is to learn to know it and, knowing it, be willing to live in it. I think one can come to know the anxiety of the unknown of death to such a point that one overcomes the anxiety as 'bad fear' and can embrace it with a kind of 'good fear.' I have witnessed it many, many times. I saw it with Rick Fairbanks."

As the dean at the small college where I met David Saetre, Rick Fairbanks gave me my first teaching job. I was in his office many times, enough to notice the dramatic weight loss one spring, the kind of loss that makes you wonder if something is terribly wrong. Oh, someone told me, he's really been into cross-country skiing. Oh, I remember thinking, that's fantastic. But it wasn't, and some six months after being diagnosed, he was dead.

"He was a trained philosopher and he could be a son of a bitch," Saetre says, "but as he was dying his reflections were really profound. At some point he stopped reading philosophy and, while he was still able to read, he started reading novels that he loved. He said, 'I just want to read stories about the endless possibility and variety of being human.' When he couldn't read anymore and became bed-bound, he would want you to tell him stories."

The story that moved around town in the days after his death is this: On the day Rick Fairbanks died, he asked his daughter to tell him a story about paddling. And so she created a narrative of a paddle that they had done together on Lake Superior, where the sea had started calm and easy and then got rough. And she said to him, "It's kind of like right now. The seas are rough." He was saying almost no words at that time, but he said, "I kind of like it that way."

Saetre says, "I know that he was fully conscious of the double meaning of his comment. Here was someone who gave a lot of thought to dying. He dove into it. He was coming to know death and wasn't in denial of the kind of fear that is attendant to dying but because he knew death, he could live in it. It wasn't that there is NO fear, but in KNOWing the possibility of death that was soon to come and of knowing his fear, he could live in it in such a way that it no longer controlled him. I think when we talk about fear of the dark, or our fear of dying, it really is fear not only of the unknown but fear of losing control."

Still, Saetre says, if we avoid all circumstances of the dark, then we live in that false world of controlling and manipulating nature so that we don't have to experience fear. "But you will be afraid in the dark," he explains. "Don't run from it. Know the fear and know the dark, and—getting back to Rick—I think that is what he was experiencing the night he died: 'I am no longer afraid of dying,' in other words, 'I am no longer resisting death. Do I know what this is going to be? No, this is going to be a hell of a ride, probably—but I can live in that, and I can die in that.' "

Then, looking at me, Saetre smiles, "If I had been quicker-witted when you said you don't think you know much about death, I would have said, well, tell me what you know about darkness, because they are deeply linked."

3

Come Together

A system of conservation based solely on economic self-interest is hopelessly lopsided. It tends to ignore, and thus eventually to eliminate, many elements in the land community that lack commercial value, but that are (as far as we know) essential to its healthy functioning. It assumes, falsely, I think, that the economic parts of the biotic clock will function without the uneconomic parts.

— Aldo Leopold (1949)

The Isle of Sark rises abruptly from the English Channel, three-hundred-foot cliffs topped by dark hedgerow lines and sloping checkerboard greens, looking as if chipped from England and floated out to sea. But that is Sark in daylight; at night, in the dark, Sark nearly disappears. With no streetlights, no cars or trucks, no petrol stations lit to daylight, just the pubs, farms, and homes of its six hundred residents, Sark emits almost no light of its own. Seventy miles south of England and just half that north of France, Sark itself covers only two square miles, but it soon may have an impact beyond its size as the world's first International Dark Sky Island.

Until about a year ago, I had never heard of Sark. My guess is most of the world's nearly seven billion people could say the same. But at least a few more know about this tiny island now, thanks to its recognition in 2010 by the International Dark-Sky Association. The IDA began its International Dark Sky Places program in 2001 with its designation of Flagstaff, Arizona, as the world's first International Dark Sky City. That category—Dark Sky City—has since been changed to Dark Sky Community, and has been joined by such designations as Dark Sky Parks and Dark Sky Reserves. And not that the IDA has cornered the market on such designations, as similar programs exist elsewhere. In Canada, the Royal Astronomy Society has its own system of Dark Sky Preserves, for example, and the United Nations Educational, Scientific and Cultural Organization (UNESCO) has initiated its own Starlight Reserves program, too. Although each varies slightly in its approach, the different programs are working toward the same general goal: protecting darkness in a world of ever increasing artificial light.

What makes Sark especially compelling is that people actually live there, with their fears of the dark, their concerns for safety, their desire for "progress." As important as it is to protect areas of wild pristine sky, it's the protection of darkness in places where people actually live that will ultimately change attitudes toward light and darkness.

"If you only want to slap patches on very dark places, you can do that to your heart's content, cover the world with dark sky parks," says Steve Owens, the Scotsman who helped walk Sark through a two-year process with the IDA. "But it wouldn't affect one single light. Whereas Sark, they had to do some light work." By "light work," Owens means that in order to qualify for the IDA recognition, the community of Sark had to take action—inventory its existing lights, change those lights that were causing excessive glare and sky glow, and promise that any new lights would conform to anti–light pollution regulations. In doing so, they met the IDA's definition of an International Dark Sky Community: "a town, city, municipality, or other legally organized community that has shown exceptional dedication to the preservation of the night sky through the implementation and enforcement of quality lighting codes, dark sky education, and citizen support of dark skies."

"They actually want the places that are on the borderline," Owens explains of the IDA, "the places that should be good but aren't, and can get good by doing some work on the lighting. They're not so much interested in places that are already exemplary, because it doesn't achieve their goal of improving lighting." A Dark Sky Community, then, acts as an example to help people understand that darkness—and good lighting—isn't something just for National Parks or communities near observatories, but something to which everyday communities could aspire.

Born and raised in Inverness, Scotland, at the edge of the famed Loch Ness, Steve Owens grew up interested in astronomy, ran a science show theater "where people blow stuff up and set fire to things," and now makes part of his living helping communities develop their dark sky identities. His first success came when the IDA declared Galloway Forest Park in southwestern Scotland the first Dark Sky Park in Europe. Galloway, which claims Bortle 2 skies, is the first in what Owens hopes will be a long list of parks in the United Kingdom to become dark sky reserves. "I don't think that's excessive," he says. "Mainly because the national parks in the UK are called 'Britain's breathing spaces,' and the measurement they take of their success in that is a measurement of tranquillity. They've done study upon study of what people think tranquillity means, and always in the top three is a good, clear night sky and no light pollution."

As important as official designations may be, Owens believes that ultimately dark sky areas will only succeed if they are supported by local communities. When astronomy programs began to take off in Galloway Park, for example, and people who lived in or near the park started to hear others saying Galloway was one of the best places in Europe for stargazing, their reaction was, Owens says with a smile, " 'Oh, I didn't know that. Do I live in one of the best places in Europe for stargazing? That's quite good.' And that filtered through eventually, and people got excited.

"It's all about education," says Owens. "It's about making sure that people are aware of dark skies. Most people, up until relatively recently, weren't. I think the real sea-change, the massive step forward, has come through the Dark Sky Parks. The Galloway Forest Park might affect hundreds of thousands of people over the next few years as they visit the park. And more than that, one hundred sixty million people worldwide heard about this effort. Certainly in the UK media, that elevated light pollution to a different level."

The popularity of the dark sky places idea stems from their focus on the positive, Owens believes. "The media definitely were interested in reporting a good news story that was about environmentalism and economics and tourism and astronomy. Also, in the UK, there's just a massive momentum behind the dark sky movement, the astronomy movement. And it's not coming from preaching about people's bad lighting, it's coming out of, 'See how amazing this stuff looks when you can get good sky.' "

But tonight, the night I have come all the way to Sark—from Paris on a train to the French coastal city of St. Malo, a ferry to Guernsey, a tug to Sark, a tractor to the village center, a horse-drawn Victorian carriage down a one-lane dirt road to a bicycle to visit with island resident Annie Dachinger until midnight—the sky is filled with clouds, and I haven't seen a single star.

"You might still," she laughs. "You should have consulted a good witch before you came." Sark has a bad habit, she explains, of toying with its visitors. "It can be rainy and drizzly all day, and just as people are getting on their boat to go back, the sun will come out. And I'll think, Oh, you bitch! How cruel!"

A hand-painted sign on the gate outside Annie's small house reads, "The Witch Is In." She says one of the carriage drivers likes to bring tourists by and tell them, "It's me mother-in-law lives there." In her sixties, with sandy brown hair and a raspy voice, she snaps her cigarette lighter open and shut while we talk. "Can I offer you anything?" she says. "Coffee, tea, whiskey?" She's part of the small group of Sark residents who made the push for the dark sky designation. "The stars here are amazing," she says as we look out her picture window, the only lights in the dim room two white candles burning. "The other night, it was like Van Gogh's *Night Café*. And I don't know—maybe I'd had a bit to drink—they all looked as if they were huge and glowing. I had to hang on the side of the house because they were making me dizzy." She laughs her raspy laugh. "The best thing to do here on a really good night is go and find a field somewhere and just lie on your back and look up. And first of all you'll see three hundred or four hundred stars, and then the more you look, the more you see, until the whole sky fills with them."

She came here from London in the 1970s and found a darkness unlike she'd ever known. "When I first got here, I thought, this is like going back five hundred years. And it is—the blackness is like velvet here. But it's an embracing kind of darkness—it's not at all scary. It's like being asleep while you're awake."

With no cars or trucks on the island—only tractors for farmers during the day—darkness brings a hush to the fields and carriage lanes. Annie says she will wake up and wonder, What's that noise?—only to realize it was the sound of her eyelashes brushing against the sheet. "Because it's so dark," she explains, "you can actually hear little tiny sounds like that. It's wonderful. You get true rest. You wake up with the sun. It makes you much more acutely aware of your own pulse, your own life flow."

She loves Sark, she says again and again. "It's such a safe place to be. As a woman, I can go out to a concert on the island, and I can walk home at midnight all by myself, a mile and a half across the island and I don't worry at all. If the moon is shining, I just walk in the moonlight. Otherwise, my trusty old torch."

Of being a witch, she says, "I am what I am, I do what I do. A witch is a wise woman, literally. Historically, they were the healers, the midwives, the people that actually looked after communities. It's an ancient earth religion, pantheistic, lots of gods. I can go out here into my garden at midnight and have a little say to whatever I want to. I can walk out star-clad, naked."

Annie finds that watching the starry Sark sky gives her perspective. "You think, What am I? I'm a flea on an enormous animal. It really does put you in your place. But we are arrogant, we are short-sighted, we don't consider our own future, because we don't consider anything else's future either. It makes me wither inside when I think that I'm a human being and I'm part of that process. It's like a sculpture that's going wrong. Who are the gods that are sculpting us? They're making a monster of us, really."

Then she laughs again. "Don't worry, Paul. That's just me being witchy now. I get a bit like that this time of night."

After talking with Annie, I navigate back to my cottage on my ancient bicycle, down single lanes through the hedgerows, wind whipping here and there, darkness all around. The larger part of the island where Annie lives is called Big Sark, and my bed tonight is on Little Sark. To get there I need to cross "La Coupée," a narrow strip of land high above the rocky coast, on a nine-foot-wide pathway built by German prisoners of war in 1945. Before railings were built at the turn of the century, to avoid being blown 250 feet to the rocky surf on windy nights like this, children made the crossing on their hands and knees.

Back to the cottage I drop my bike, struck at first as much with the lights of Guernsey, ten miles east, as with the dark of Sark. But when I walk into a nearby hillside field, the glare from Guernsey blocked by the slope, I realize what makes Sark unique: While the Sark sky is impressively dark, the land is even darker. The ocean surf, the whirl of wind, the baaa of sheep out in this field—I can hear all this but can see only the shapes of sleeping cottages, no lights in or out, and, where the roofs end, the stars begin. For, just as Annie predicted, the sky has begun to clear on the horizon around me.

These are maybe the most exciting stars, those just above where sky meets land and ocean, because we so seldom see them, blocked as they usually are by atmosphere or—especially these days—by pollution. The single-lane gravel roads between hedgerows, the sleeping horses in the barns. This island takes you back in time—not only through the absence of cars and trucks but because of these stars at the edges of Earth.

And, as I grow more and more accustomed to the dark, I realize that what I thought were still clouds straight overhead aren't clearing and aren't going to clear, because these are clouds of stars, the Milky Way come to join me. There's the primal recognition, my soul saying, Yes, I remember. With cliffs all around I feel as if I am on a pedestal set among the stars.

Tomorrow I will head to Guernsey, a bobbing diesel-churned journey, and find cobrahead fixtures, unshielded lights, the insistent roar of the motors that rule our lives. But tonight in a field on Sark, I lie staring up—and around—at the starry sky, a man on his back in a field, all but disappeared.

When Aldo Leopold wrote *A Sand County Almanac*, he placed his "land ethic"—his argument that we ought to treat the rest of creation as ethically as we hope to treat other humans—at the heart of his book, and at the heart of the land ethic he placed the idea of community. Leopold believed the reason humans are so shortsighted in our treatment of the natural world is that we do not see ourselves as part of a community with it. He argued that, while we have made great strides over the centuries toward expanding our notion of the human community to include a wider range of race, gender, and ethnicity, we have not made the same adjustment for the land. "All ethics so far evolved rest upon a single premise: that the individual is a member of a community of interdependent parts," he wrote. "The land ethic simply enlarges the boundaries of the community to include soils, waters, plants, and animals, or collectively: the land." Leopold believed it was not enough to only value those members of a community who provided obvious economic value, such as deer or pine trees, because "most members of the land community have no economic value" or no easily defined value. Instead, Leopold argued for valuing the whole—that every member of the community is valuable, whether we understand that value yet or not—and for acting accordingly. "Examine each question in terms of what is ethically and aesthetically right, as well as what is economically expedient," he advised. "A thing is right when it tends to preserve the integrity, stability, and beauty of the biotic community. It is wrong when it tends to do otherwise."

Working in the desert Southwest during the first decades of the century, Leopold would have known fantastic darkness. Even when he moved to Wisconsin in 1924 he would have known real night while at "the shack," his retreat forty miles outside Madison. Darkness doesn't show up explicitly in his writing, but Leopold would have understood the costs of losing it. To value darkness would be to follow his desire for us to enlarge the boundaries of our community. Ecologically this is vital: If we truly value nocturnal and crepuscular creatures, for example, then we won't allow our artificial lights to destroy their habitat. Leopold's thinking applies too in that the values of darkness aren't always economically obvious. How do we quantify the value of the darkness that provides passage for migrating sea turtles or shorebirds? Or the darkness that hosts the starry sky that might inspire the next Van Gogh?

We have not yet learned to think about artificial light as Leopold would have us do. That is, as an ethical choice: Do we care that our lights shine into our neighbor's bedroom? Do we care that our lights dilute the darkness on which bats and moths and migrating birds depend? Do we continue to ask more and more predominantly minority citizens to work through the night, knowing full well the health risks they assume? My sense is that we so take for granted electric light that, not only do we forget how amazing it is or how beautiful it can be, but we are unaware how our use of light affects the rest of our community.

This is not what I expected to find.

In almost every way, it's better. The mountainous country covered with a thick forest of sugar maple and yellow birch, fir, and pine, crisscrossed by hiking trails and haunted by creatures as ephemeral as luna moths and as regal as moose. But in one significant way, it's worse: There are no stars here. There is incredible darkness—when I step from the observatory with my host I cannot see my hand before my face, and even after talking together for twenty minutes in that dark, he is only a vague shape three feet away. Unfortunately, Mont-Mégantic Starry Sky Reserve lies completely socked in with fog, rain, clouds thick as wool. I'll not see any stars while I'm here, but I will leave thinking that, in some ways, Mont-Mégantic is the most impressive place I've yet been.

Located in southern Quebec, just across the border from Maine, Mont-Mégantic National Park (Parc National du Mont-Mégantic) is home to the IDA's first Dark Sky Reserve, created in 2008. The IDA calls its "reserve" designation "the epitome of IDA's mission" and describes a Dark Sky Reserve thus: "working to preserve a central core that is valuable because of its natural light, communities band together to create public awareness campaigns and conduct retrofits to restore the night sky." The definition sounds as if it were written to conform to what Mont-Mégantic has done, rather than the other way around. Mont-Mégantic has made itself a model for future initiatives to protect darkness and night skies while meeting the needs of twenty-first-century human communities.

It's like a whole different country up here. Coming north from the States, the change is immediate—everyone speaks French; even the road signs are in French. Of course, I knew this would be the case, and I find it a pleasure, but the difference in language belies an independent streak that applies to the reserve as well. For one thing, the sign as you enter does not read "Dark Sky Reserve," but "Réserve internationale de ciel étoilé"—International Starry Sky Reserve, which the MM folks think sounds more positive. More significantly, they are doing things here that almost no others are doing. Over the course of less than a decade, Mont-Mégantic has managed to enlist the support of more than sixteen different local communities for their dark sky efforts, put into place laws governing lighting, replace more than three thousand lighting fixtures in those communities, and introduce the concepts of light pollution and dark skies to more than five hundred thousand visitors. As a result, despite lying only a hundred miles east of Montreal (the second largest city in Canada and the seventh largest in all of North America), the skies over Mont-Mégantic remain Bortle Class 3 dark.

The Starry Sky Reserve consists of a scientific observatory (built in 1978, and still the largest and most powerful on the eastern seaboard of North America); a "popular" observatory, built in 1998; and the ASTROLab, built in 1996, which features displays, movies, guided talks, and tours. As we stand talking outside the popular observatory, my host, Bernard Malenfant, tells me that when he arrived here thirty-three years ago, he needed to carry a flashlight to work on the outside edges of the observatory, but twenty years later, he no longer needed the help—the light pollution in the skies overhead had more than doubled. Now, after the success of the past few years, "the sky is nicer here than it was in 1978. And, because it's a regulation, a municipal law, people just can't put up lights anymore. In two hundred years we might be the last place where it's still dark. Hopefully not. There will always be the Atacama Desert in Chile, but a place where people are living and the sky is still dark? It's a long-term project, but one of our goals is to preserve the sky for our children's children."

Much of the credit for this success goes to the self-effacing and jocular Malenfant. Though he claims to be merely the night custodian, his role as chairman and founder of the ASTROLab has gone far beyond such basic work: Malenfant has had a hand in making each of the reserve's buildings become a reality. In fact, the ASTROLab, which now hosts packed shows throughout the summer, was his idea. He says he realized the need after seeing how many visitors were "driving five hours from Quebec City, looking through the telescope for five minutes, and driving back home." The Starry Sky Reserve now features a dark sky festival in July and a meteor festival in August and welcomes tens of thousands of visitors every year. It's this public outreach that makes Mont-Mégantic unique. Yes, it's dark here in part for the scientific observatory, which is owned collectively by two universities. But many dark sky places can claim that fact. What's different about Mont-Mégantic is that, while many other dark sky locations focus first on an observatory, with any benefits to the public secondary, here that feels reversed.

In truth, major astronomical observatories can be pretty boring to visit. They are often tough to reach and have limited, if any, public visiting hours. Even if there is a public telescope, the experience can too often be as Malenfant described: a long drive there, a quick look, a long drive home. But every year, Mont-Mégantic hires a crew of college students, many of them astronomy majors or simply young people with excitement for the stars, to serve as guides at the ASTROLab and popular observatory. The effect is that every visitor has the opportunity for personal interaction with someone who's not only knowledgeable about the universe but enthusiastic as well. While Mont-Mégantic has made its displays and its movies both entertaining and enlightening, it's this personal connection that draws so many visitors each year—and then draws them back. For many of the visitors, especially those from the cities, a visit to Mont-Mégantic is an opportunity to experience what was for thousands of years the common human experience of being together with other people under a starry sky. For the guides, too, Malenfant tells me; many come back every summer even if they have moved on to other jobs, simply to enjoy the experience of being with others under the Milky Way, or around a campfire exchanging stories and songs through the night.

As I listen to Malenfant describe these nighttime gatherings, I think of how darkness brings us together with those we love. So many of our most intimate, romantic, memorable experiences—a campfire in the woods, a candlelit dinner, time spent in a bedroom with a lover—are experiences we illuminate with flame or moonlight, with subtlety. During the day, we wear the bright light of the sun, see ourselves in the mirror, imagine what others think, and shy from revealing our thoughts, our body, our fears. But darkness allows us to lower our defenses—we can say what we want, do what we want. We have the opportunity to rely on other senses, on touch and taste and hearing. By providing the context for intimate light, darkness brings us closer.

On Christmas Eve at the candlelight service in the downtown Minneapolis Lutheran church I grew up attending, near the end of the service the electric lights are drawn down and a flame passes from one person to another, candle to candle, until the entire sanctuary is lit only by candlelight. I remember once as a child noticing a blind man in the row ahead of us, holding his candle close to his face with both hands to feel the flame, his eyes closed, his smile.

The small red Christmas lights my mother wraps on the tree each year, the maple-scented candle burning on my desk as I write, the fireplaces and campfires and moonlight I remember from years gone by, and the possibility of such experiences in years to come—it's as though with the flick of a switch, a room's sudden bright, all are erased.

The emphasis at Mont-Mégantic on connecting the public to darkness took a big step forward with the hiring, in 2003, of Chloé

Legris as outreach coordinator. Originally hired on a six-month contract, Legris stayed five years, and for her efforts was named Scientist of the Year in 2007 by Radio-Canada. Trained as an engineer and blessed with natural charisma, Legris worked tirelessly to connect the goals of the park with the realities of the local communities. Remarkably, when she began her job she knew next to nothing about dark skies. But as she began to learn, she says, she "fell in love with the project." She also quickly realized that she would have to work hard to be taken seriously. "I was not emotional about stars and dark skies," she says. "I approached it in a pragmatic way. If I were an electrician, what would I care about? I wasn't trying to sell to people, I just told them it was logical. Pragmatic people don't want to hear about the problem, only how to solve it. The electricians told me, 'Don't talk to me about astronomers. I like to fish and I like to see the stars, so how do we do this?'" Over the course of nearly six years, Legris went "everywhere" in the area surrounding Mont-Mégantic, meeting with politicians and business leaders, conducting training sessions on good lighting, urging communities to adopt lighting regulations, and raising money to pay for the retrofitting of existing lights. "We took care of everything," she says. Because the power company Hydro-Québec had a federal mandate to become more energy-efficient, the company had set up a fund for new approaches to using less energy, and changing lights and light fixtures fit the bill. A significant part of Legris' work was simply helping local people to appreciate the breathtaking skies they see every night. "It's teaching people we have something special here," she says. "For them, it's just normal. What do they want to see? They want to see a McDonald's in their town. I'm joking, but it's not so far from the truth," she laughs. Nonetheless, her efforts began to sink in, sometimes in amusing ways. The mayor of Notre-Dame-des-Bois, a tiny community at the foot of the mountain, told her that his mind changed when he was driving one night in the country. "It was really funny," she says. "He stopped on the road to pee, and he rose his eyes in the sky and he thought, It's true, it's so beautiful what we have. And I didn't realize how much I appreciate it because I see it all the time."

Legris now lives and works as an engineer in nearby Sherbrooke, at more than one hundred fifty thousand people the largest city close to the park, and she can see the results of her work in the lighting fixtures around her. To hear her say it, however, the credit for "her work" really goes to others. When I ask her what she is most proud of, she says, "the engagement of the people behind this. Partnering with me. There was so much support in the community, with so many people. And I'm proud of these people for being proactive, and deciding to go forward. I couldn't have done this by myself. So really it's a community success. Whenever I had a chance to give a presentation and explain to them about the situation, most people were like, 'What do we do?'

"I found the job very inspiring," she says. "It wasn't just building a bridge, it was protecting the ability to see the universe."

As education director for Mont-Mégantic National Park and scientific coordinator for the ASTROLab, Sébastien Giguère describes his job as communicating to the public the mission of the Starry Sky Reserve. "They're doing the science," he says of the scientific observatory, "and we're sharing science. And science isn't only about equations and white coats, it's about the mystery of our presence in the universe and why are we here and why this is so fabulous." Giguère has a pleasant nature and conveys a sense that he's committed, serious, and sincere. "And I have a strong wonderment muscle," he laughs. "I don't know if we can say that in English, but I like to say that to guides here. We're not here to make a college course, we're here to make people amazed about nature, so we have to see it in your eyes, and you have to be expressing this sense of wonder about everything. I like to cite Einstein, 'He who can no longer pause to wonder and stand rapt in awe, is as good as dead: his eyes are closed.' "

Giguère tells me that he and a number of the guides traveled to New York City to visit the Hayden Planetarium, which they enjoyed (and wished for "a fraction of their budget"), but that "one thing that struck everybody here was that they have so much technology but nobody there to share the amazement, the wonder, no one to talk with about this. And that's not just in New York; in the Montreal Science Centre it's the same. I like what Stephen Jay Gould said: 'We will not fight to save what we do not love.' I always remember that our link to nature is not just cerebral, it's also emotional."

What impresses me most about Giguère—and this reflects so much of what I've learned at Mont-Mégantic—is that while committed to the scientific mission of the observatory, his dedication to darkness goes beyond simply protecting the view of the sky for astronomers.

"The first mission of the project—what is now the Starry Sky Reserve—was to save the scientific viability of the project here. But I like to say that maybe in a few decades what we will realize is that the most important heritage of this project is to have preserved the possibility of the fundamental experience of the night sky.

"A lot of people come from the cities, and they are amazed, they don't believe it, they don't remember how starry a sky can be. People just sit down, outside here, because they have vertigo. Compare that to how one of my colleagues came back from China and told us that in a lot of places there is so much pollution in the air that only one percent of sunlight reaches the ground. And so, you know, I am crying because we are not seeing the stars at night, but there you can't even see our own star in the daytime. What is happening to humanity when children grow up without seeing the sun?"

Giguère seems like someone for whom Leopold's statement about the consequences of gaining "an ecological education" would make sense.

"That's a great question, how to be positive without being naïve.

I feel a kind of natural amazement toward the world, and even if I'm depressed witnessing what is happening right now, it just can't kill this amazement. I also feel lucky working in a place where nature is everywhere, and I have so many opportunities to be amazed and love nature, because with the starry skies, with the mountains, with the lakes, birds, and animals, I would find it hard to go back into big cities. I'm not used to them anymore, being stuck in the traffic and all the pollution, and all the stores and streets and pavement everywhere.

"So I try to do my part, doing some talks, but I still feel that it's not enough. My girlfriend tells me not to put too much on my shoulders. But when we know what is at stake, that it's the first time in history our impact is so wide. And then, that a lot of people don't notice . . . ," he says quietly.

"I like to think the fact the stars are disappearing from the sky comes back to our relation with nature, and our way to inhabit the earth. And so closing our only window to the universe appears to me like a great symbol of how we are separating ourselves from nature. It's like how people don't go out of the cities and so they don't know; they are just stuck in their bubble.

"We can't see the universe anymore—it's not the most dangerous thing that we are doing right now but it's a powerful symbol."

Giguère admits with a laugh that he talks so much about the importance of wonder that people recognize him as the "wonder-keeper," because he is also the goaltender on the local hockey team. "This feeling isn't only directed toward nature but toward humanity as well. Some people think astronomy is not that important, compared to ecology, because of what's at stake. But I think the more you are aware of this incredible cosmic evolution story, the more you are aware of the miracle of life. When you know about the vastness and emptiness of space and then you look at this little pale blue dot, you can really develop a feeling of responsibility because you know how rare and miraculous and beautiful it is.

Almost every astronaut coming back from space has said that the most important thing he has lived there is just looking at the earth and realizing how precious it is and how our borders are all relative. William Anders, the astronaut who took the famous photo of earthrise, said, 'We came all this way to explore the Moon, and the most important thing is that we discovered the Earth.' "

As I drive down the mountain from the observatory, it's after midnight. My headlights can barely show the road ahead, the fog is so thick. I drive slowly on the steep, curving roadway. I came all this way and this is not what I was hoping to see at Mont-Mégantic. I was hoping for a starry night, to see how the darkness here compares with Sark, or Cape Cod, or the lake in northern Minnesota where I feel at home. But I also didn't expect to find what I have—a group of people so committed to making Mont-Mégantic what it is. And each—Malenfant, Legris, Giguère, several young guides, scientists, and upper-level administrators—plays a vital role. With the people at Mont-Mégantic I felt a kinship, and in that kinship I found at least one antidote to the sadness in Leopold's quote about living alone in a world of wounds. I found a community of people who are choosing to do everything they can to help the community around them become aware of its riches.

"I love my sky. This is my problem."

Cipriano Marin and I are sharing lunch when he says this, sharing papas arrugadas with mojo picón and mojo verde and glasses of red wine in Teide National Park in Spain's Canary Islands. It's a sentence I know I won't forget, a sentence I have heard from so many others in so many different ways. To love a sky—to love anything—is only a problem if you know that love is in danger.

Cipriano knows. A native of the Canary Islands, which lie southwest of the Spanish mainland off the coast of Morocco, he grew up with a clear, dark sky. Now in his late fifties, with salt-and-pepper hair, he has seen that dark sky gradually fade, due especially to the city lights of Las Palmas and Santa Cruz de Tenerife. He feels the loss especially hard, he says, because he grew up on the islands. "The sky for the islander is very important," he says. "We have only two natural resources, the sky and the sea. On an island, the sky is part of the landscape, part of the identity of the island." Luckily for the rest of the world, Cipriano's activism on behalf of dark skies has reached far beyond the islands he loves. For twenty years, he has worked with UNESCO, and in 2007 he helped to organize an international conference in the Canaries that resulted in a declaration "in Defence of the Night Sky and the Right to Starlight," and an innovative program called Starlight Reserves.

And so I have traveled to the Canaries to meet Cipriano, and to eat "wrinkled potatoes" with red and green salsas and to talk about protecting darkness. But I have also come here to see a world-famous night sky.

I'm talking about a night sky that leaves you breathless, that makes you want to study the stars, or write poetry, or dance. In the months just before I traveled here, taking the two-hour flight from Madrid with a plane full of European tourists on holiday, a photographer had assembled a series of time-lapse images of the sky over the Canaries, set it to music, and put it on the Web. I knew about this because just about everyone who knew about my writing this book sent me the link to these photographs, often with a note saying, essentially, *You should go here!!!!!* For a number of reasons it isn't fair to compare a photograph of the night sky with what you can see with your own eyes—a camera's ability to keep its lens open for long periods of time gathering light and to focus every pixel in the frame rather than just the center of the view, for example. But I still hoped that the sky I would see in the Canaries would be pretty close to these photos.

I'm not alone. The Canaries are home to the world's newest major telescope, which also happens to be the world's largest single-aperture optical telescope: the Gran Telescopio Canarias (GTC).

The GTC sits on the edge of a volcano, along with several other telescopes at the Roque de los Muchachos Observatory on the island of La Palma, and attracts astronomers from all over the world. It isn't only having an incredibly large telescope that makes the Canaries special, though. The Canary Islands are one of a handful of sites in the world extremely well situated for viewing the night sky.

Urban growth and sprawl—with its attendant light pollution—has rendered observatories in and near cities obsolete. So while there are observatories in places like Paris, Los Angeles, and London, few people use them. For example, the Paris observatory is a great place to visit if you appreciate history and enjoy imagining the world as it used to be—an elegant building set in fields outside town where silk-stockinged, wig-wearing viewers took in the sky—but it now lies surrounded by the concrete fields and blooms of electric light that fill the French capital.

These days an absence of light pollution is only one requirement for an excellent observatory site, and only a few locations in the world work. Especially important for optical telescopes like the GTC is turbulence (or lack thereof) in the earth's atmosphere. Because turbulence makes holding a steady image difficult, the best locations in the world are those in the earth's mid-latitudes, especially on the west coast of a continent or island, where the dominant west-to-east flow of air moves smoothly in from the sea. Good weather—few clouds and little rain—is also key, and so deserts are often excellent locations. Accessibility, stable ground (no active volcanoes), moderate altitude—when you add up the different requirements, only a handful of spots around the globe stand out, including the Canaries, the Hawaiian Islands, Baja California, northern Chile, and South Africa. Together, the modern observatories in these locations create a network of telescopes, "an ensemble of windows open to the universe" (as Cipriano says), that provide the best views of space we have from our home here on Earth.

* * *

Unless you have been raised by diplomats, you probably have not spent much time with someone like Cipriano Marin. Without it seeming corny or naïve, he talks in language that seems made for "declarations" written "in defense" of certain rights. In fact, he makes the idea that the ability to see a starry sky is a basic human right, like clean water and voting, seem wholly justified. My sense is that in the United States, a country known for thinking it has the last word on rights, most of us have never considered a view of a starry sky as a "right." But the UNESCO declaration does: "An unpolluted night sky that allows the enjoyment and contemplation of the firmament should be considered an inalienable right of humankind equivalent to all other environmental, social, and cultural rights." Cipriano admits that such a right is difficult to uphold—that while it's not hard to find philosophical support for the idea, legislative support is tougher to secure. Because, I ask, then somebody would have to do something about it?

"*Sí.*"

Still, Cipriano has done impressive work to raise the issue of dark skies to international consciousness. The report from the 2007 conference includes statements of support from the Spanish minister of the environment, the vice president of the European Parliament, and a long list of directors, presidents, and secretaries-general, followed by four hundred pages in support of a right to starlight in essays written by scientists, artists, organizers—people like Travis Longcore and Chloé Legris. Together they argue that "a view of the starlight has been and is an inspiration for all humankind, that its observation has represented an essential element in the development of all cultures and civilizations, and that throughout history, the contemplation of the firmament has sustained many of the scientific and technical developments that define progress." Cipriano calls the night sky "an essential element of our civilization and culture that we are losing at a fast pace, and whose loss would affect all countries in the world."

Maybe that's the most impressive thing, that this man from a small island in the Atlantic is doing everything he can to rally the world. It would be so easy for him to simply stay on his island and not bother. Cipriano's passion comes from his knowing that there is nowhere to run, nowhere to escape, nowhere to drive to get away from light pollution. In this, he is much like citizens of other island nations who feel the effects of rising sea levels from climate change. For most Americans, the need to address such problems seems decades away. "For the islanders the universe is very closed," he tells me. "And that is a problem but also an advantage."

Cipriano sets his wineglass on the table. "Rafael Arozarena, a writer from my homeland, an islander from the middle of the ocean, synthesized the whole spirit of the declaration in a beautiful, short poem:

My inheritance was a handful of earth
But of sky
All the universe.

One of the most meaningful things about the Starlight Reserve concept is how detailed are its dimensions, categories, criteria, and recommendations. Cipriano and the more than one hundred international experts working alongside him have put forward in a compelling and novel way their reasons for the need for Starlight Reserves and their vision of what those reserves would be. Rather than simply assume that all protected areas are protected for the same reasons, the different Starlight Reserves imagine several types of these areas: Starlight Natural Sites safeguard nocturnal habitats; Starlight Astronomy Sites protect our view of the stars; Starlight Heritage Sites preserve "archaeological and cultural sites or monuments created by man as an expression of its relationship with the firmament." Starlight Landscapes preserve "natural and cultural landscapes related to starlight where natural manifestations or human works beautifully blend with the view of the firmament." And, finally, Starlight Oases/Human Habitats are dedicated to the protection of darkness in areas that include rural communities and small villages.

Cipriano believes Starlight Reserves have great, as yet untapped, potential for tourism. He names especially those World Heritage Sites all over the world, most now closed at night, that could be developed for night tourism to the benefit of local communities. "The night is missing in the tourist destinations, especially the ecotourism destinations," he explains. "It's important to offer the night." Through such night tourism he believes these Starlight Reserves present opportunities for communities around the world to develop in ways that protect darkness but allow people to enjoy modern notions of progress. "We need to relate the idea of seeing the stars to the advance of modernity," he tells me. That is, those who care about darkness should look for ways to protect the night through supporting economic development, rather than simply hoping there will always be undeveloped areas to serve as refuges of real night. "If you see the satellite image of North Korea, it's dark," he explains, "but that's not a good solution."

The Korean image is one of the most dramatic nighttime views we have of the earth from space. On the Korean peninsula, South Korea blazes away like any developed country, and Seoul like any major city. But just north of that enormous city, a sudden line of darkness marks the DMZ and begins the expanse of darkness rising north the length of the peninsula. This is long-suffering North Korea. The view of such sudden darkness back-to-back with that of light-tattered South Korea is dramatic and, in some ways, appealing, but no one would wish the lives North Koreans endure on anyone. Of course, when you look at satellite photos of the world at night, there are many other areas of the world with large human

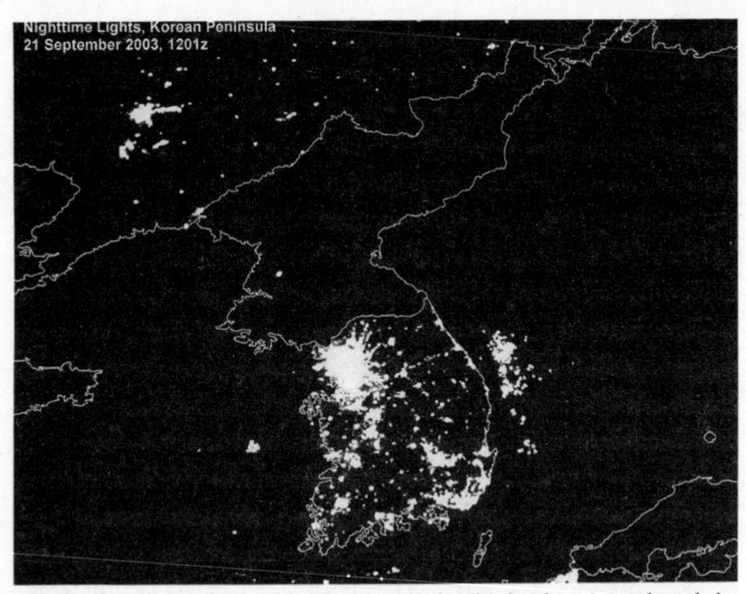

The Korean peninsula at night showing the bright developed South and the dark undeveloped North. *(NASA, DMSP)*

populations that are still dark at night—great swaths of sub-Saharan Africa, for example, as well as large patches of Asia and South America. While it would be wrong to deny people in these parts of the globe the gift of artificial light at night (and fascinating programs are already bringing solar lanterns to far-flung populations, for example), Cipriano and many others hope that light can spread without the associated costs we in the West now suffer. The hope is for a different style of progress, with the benefits of the modern world reaching more and more of its citizens but with the map of the world at night growing darker and darker.

You definitely should not eat a huge breakfast before driving up the winding road to the GTC observatory. I have found this out too late, and I am remembering when I was thirteen and traveled a similar road outside Mexico City with my traveling baseball team in a bus with large Plexiglas windows that opened only at the top, a bus that would not stop for sick American boys. Luckily, today my discomfort doesn't develop into needing Cipriano to stop his old Mercedes, and finally we emerge from the lush forest of lower elevations into the higher volcanic landscape that is home to the observatory.

The mountain is actually home to several telescopes, including a solar telescope trained on the sun (the surface of which looks like orange juice set at roiling boil), radio telescopes that in their oval shape remind me of a human eye ("They are," Cipriano says. Eyes, that is, to look out into space), and optical telescopes owned by the Dutch government, by the Japanese, and by the Italians ("Telescopio Nazionale Galileo," says one sign). Do we still imagine an astronomer at the telescope, if not in robes with a wizard's cap, then perhaps sitting high up in a chair, peering directly into the eyepiece? Those days are past, to the point that astronomers here refer to being personally present to view the cosmos through the telescope as "classic mode." Most of the viewing done here (80 percent) is by astronomers sitting at their computers in their labs back home. They schedule a time and a location in the sky, and the observatory sets the telescopes to their preference. That doesn't make these telescopes any less impressive to me. I stand next to them, thinking that this is the edge of the earth and these are the earth's best eyes: There's arguably nowhere else in the world where we can see farther into space than with the GTC.

Standing underneath the GTC—we're allowed into the dome, thanks to the fact that Cipriano knows everyone—is like standing under a cat's cradle made of steel, an enormous spiderweb of incredible weight. Looking around at the dome's silver walls, I think how moving it must be to see the sliding doors open and have the universe exposed and framed, a most wonderful window. I imagine the frame and the slow revealing would make the sight of the sky even more dramatic, as looking up through the stained glass windows of a cathedral might inspire a view of heaven.

"It's the same vision," says Cipriano. "A cathedral to connect the man with the God. Sometimes people say here is like a monastery, because you are in a very isolated place, and you are contemplating the sky."

Unfortunately, tonight will not be a good time to contemplate the sky. None of the three nights I'm in the Canaries will be. An enormous sandstorm has arisen, a *calima*, a hot, stifling wind from the Sahara desert blown west to muddy the Canary Island skies. When I ask one of the astronomers at the observatory if he will be observing tonight, he grimaces. "Oh, no, it's horrible out there."

There's nothing I can do. There won't be any comparing of my experience seeing the night sky here with those Internet photos. No comparing of the world-famous Canary Island skies with my memory of that night not so far from here in Morocco, when my first thought was of swirling snow. No new experience of a night sky I would never forget. Cipriano, who, from the moment we met at the airport, has been a warm and generous host, is visibly disappointed. "It is not like this here," he says. "*Calima* in this time of year is very strange."

At first, I'm disappointed, too. I was looking forward to seeing an amazing night sky. But just as both Sark and Mont-Mégantic offered the unexpected, so my time with Cipriano does as well. Rather than images of breathtaking starry night, I'm taking from the Canaries images of the future. Because of the work being done right now by people like Cipriano, there are still dark places on every map. Whether through Starlight Reserves, night tourism, or some other idea he hasn't yet discovered, Cipriano Marin is doing all he can to help the rest of us realize we have a right to starlight and—perhaps more importantly—why we ought to claim that right.

But there's this, too: My sense is that Cipriano isn't as concerned with his own right to starlight as he is with the rights of others, especially those who haven't yet been born. Contemplation of a starry night sky—"our common and universal heritage," he calls it—"is increasingly difficult to the point that it is becoming unknown for new generations." I'm not going to see the stars in the Canaries on this visit, but I could come back next week, next month, next year, and see them then. That's my right—that's a possibility for me, for us. But if we fail to act now to protect and restore the skies we still have today, we will take that right from future generations without their even knowing what they have missed.

"The big problem of the new generations is if you never know the grandeur of the sky, it's impossible to reclaim it," he says. "And finally, I think this is the most important motivation for me."

In other words, "I love my sky. This is my problem."

A week after standing under the world's newest optical telescope, I'm in the Museo Galileo, in Florence, Italy, in front of the world's oldest—the last two in existence known to have been made by Galileo Galilei, in 1609–1610. The longer of the two is tan, almost bamboo in appearance, and the shorter one is the darkest golden brown. They look fragile, and it's probably a good thing that they are behind a wall of thick glass; it looks as if you could easily crack them over your knee. By today's standards they are like a child's instrument, but four hundred years ago they were state-of-the-art, and Galileo spent his nights in Padua, Pisa, and Florence seeing what almost no one else on earth could see. As the astronomer Tyler Nordgren told me, "Four hundred years ago, everyone in Florence could see the stars, but only Galileo had a telescope. Now everyone has a telescope but no one can see the stars."

I linger near Galileo's telescopes, then round the corner and stand transfixed: I did not expect this—a dark, cool room full of globes of the night sky from the seventeenth, eighteenth, and nineteenth centuries. *Globo celeste*, they are called in Italian: "celestial globe," maps of the night sky. Several are enormous, four or five or even six feet in diameter. Crafted of wood and lacquer, they shine an aged mahogany brown. Made with the best available astronomical knowledge of the day, they reflect in their constellations the shapes by which people made sense of the sky. And such shapes! A massive globe from 1693 by Vincenzo Coronelli catches my eye—a giant lion, a huge bear, and a long, curving snake are painted on in the shapes of constellations. Imagine looking into the sky and seeing such a menagerie of wildness crawling, swimming, and flying overhead. Unlike our contemporary skies washed flat by light pollution, these were three-dimensional skies—the detailed depths of space obvious to any sky-watcher on any clear night—and the creatures depicted have sinuous muscles and delicate eyes. You can't touch these globes, can't spin them around and stick your finger where you'd want to live or travel next. They rest, stilled on display—so you orbit them instead.

The men who made these globes thought that the stars were unmoving and unchanging; that the sky revolved around the earth each night. Now we know "better," but along with what we know comes the knowledge of what we have lost, since these globes display a sky most of us will never see. But then, they come from a world that no longer exists, in which creatures like this were more numerous and more widespread.

Walking this room, I imagine Coronelli here, asked to build a contemporary *globo celeste*, and his perplexed, confused face. The result? Perhaps a palomino globe of gray-black, gray, and splotches of white. Would the animals still be there, the enormous lions and tigers and bears, the great-winged birds and long snakes with their grinning eyes? I bet not. The globes with shiny black backgrounds and storybook animals, with clear lacquer covering a wild sky, those maps are gone; he wouldn't even try.

But I imagine him making another *globo celeste*, this one smaller, yet still exquisitely painted, still breathtaking in detail. It's a map of the earth still flowing with creation, one you can spin, and when you stop it with your finger, there is some tiny detail—he's been working all night, every night, and sleeping during the days—some miraculous beauty, some wonderful example from each location at night. The white flower of a night-blooming saguaro cactus, the feathers from a great-horned owl, the scrunched, smiling face of a particular bat—here, I'm spinning it, I stop it in the north, where I want there to be something still—he's painted the black-and-white feathers of a loon. This would be a *globo celeste* of the present-day world, the one that remains, like the older globes but smaller, because he wants us to look closer, to see what's still there.

And even another—he's excited to be at work again—a *globo celeste* for which he's gone all out with today's technology, a type he couldn't have made back then, one that causes him to wonder what the king would have given for this: a globe of night sounds, so that by touching your location you hear the night there—the cricket song, the ocean surf, the frog mating calls. This is a globe you spin in the dark, eyes closed, one you turn and then listen to: the world at night, a geography of nocturnal sound, still there, still waiting, for our ears.

I have come to Florence to see the museum and I have made sure to come when the moon was full, so that I could walk this beautiful city in the moonlight, imagining what Galileo saw. But after talking with Cipriano about World Heritage Sites and night tourism, I am also eager to see how well Florence—the historic center of which was named a World Heritage Site in 1982—offers the night.

The short answer is that it doesn't. While the center of the city may be historic, the modern city has a bursting population of more than 375,000, with more than 1,500,000 in the greater metropolitan area. This is a major Italian city engulfed by artificial light, and not a small town surrounded by darkness. But what's unfortunate is that even within the historic center, lights are allowed to clutter and glare as much as they do in any like-sized city. When I look down to write in my notebook, I can barely see the page from all the blue dots in my eyes. The fact of the city's beauty—when I first see the Duomo I say "*Wow*" out loud—the fact of its being a UNESCO World Heritage Site, seems to have little bearing on its presentation at night. There seems to be little more than perfunctory attention paid to creating a unique nocturnal ambience. As I walk the city's historic stones, the moon relegated to irrelevance by the city's flood of light, I think over and over and over, What an opportunity missed. How much more beautiful would this city be with no glare, no blazing bright lights; if it were instead lit with care and thought, perhaps even lit with candles and moonlight. How beautiful would Florence be were the moon allowed back into the city, to wash its light on the Renaissance towers, the stone walls and courtyards, the open squares and narrow streets?

And I do find streets here and there without glare. On some, you can see for blocks to the end of the street—so much more thoughtfully lit, so much more welcoming. In fact, these streets are so clearly more inviting that I immediately think, Someday this will be the standard. Someday this is what people will expect and demand. Someday people will grow sick of bright, unnecessary light. But as it is, it's as though Florence fights itself. That perhaps it employs two lighting designers, one with a philosophy of beauty and the other with a philosophy of fear. For that is the excuse the current politicians use—that with all these tourists (to say nothing of citizens), we need this light for safety. But especially in a place like Florence, that argument feels silly. The streets are full of people—even after midnight on a Sunday night, there are plenty of couples, groups of friends, happy to be out walking (though I hear as much English spoken as anything, certainly more than Italian). Are we so afraid of the dark that we must sacrifice what would be such an experience of beauty, the human-made beauty of the buildings and streets, flames and fires, with those beauties we've no hand in making—the moon and the stars and the darkness?

While at the museum, I had met Karen, an American married

to an Italian and working as a docent. She told me that the lighting level in the city has increased a tremendous amount in the last decade; that because her little street is lit more brightly now, people use it as a toilet, urinating on her apartment building under her window. She and her husband no longer walk after dark, she said, "because I don't want him to get angry and argue with someone who has a knife." She explained, too, that Florence has something each year they call Notte Bianca, or White Night: The museums and stores stay open much later, some through the night, and the lights are left on. "It doesn't have that much to do with appreciating darkness," she admitted.

As I walk back to my hotel after midnight, I think about how we go inside and the beautiful city—just as beautiful at night—goes unnoticed, except maybe by the garbage men, the Carabinieri in their squad cars, the gelato girls biking home after a long day of serving Americans. I think of how in Belgium now, and in France, one day a year is declared "the day of the night," and cities and villages across each country spend the night without lights. People come out to join in activities celebrating night, to raise awareness of energy consumption, light pollution, and the beauty of darkness. Activists I've spoken to in Paris hope the movement spreads across the continent.

And that's the thought I have before going inside, leaving the moon above the city. What would that look like, a European day of the night? What would it look like from space; what would it look like on the streets? How would it look here in Florence? Where would it lead?

2
The Maps of Possibility

To speak about sparing anything because it is beautiful is to waste one's breath and incur ridicule in the bargain. The aesthetic sense—the power to enjoy through the eye, the ear, and the imagination—is just as important a factor in the scheme of human happiness as the corporeal sense of eating and drinking; but there has never been a time when the world would admit it.

—John C. Van Dyke (1901)

Before imagining the night we might know in the cities and towns where we live, I wanted to return to one of the darkest places I have ever been. From the northwest corner of Nevada up into the eastern Oregon deserts spreads one of the last great areas of natural darkness left in the United States. And here I am, in the Black Rock Desert, with one good friend, two fold-up chairs, and the last few minutes of twilight.

I have been here before—I remember waking before dawn one night to find a blood-red waning crescent moon lifted barely above the eastern horizon, and I remember another night of unsettling wind—but I've not been here looking for darkness. The drive from Reno is barely two hours, taking us out past Pyramid Lake on Highway 55 to Gerlach, the last town before the desert. When you leave town on the paved highway, curving at the base of hills, the desert playa begins to expand off to the right, and before long you simply turn off the highway and begin your dust-raising race across the roadless land. It can be unnerving at first, having no road—car tracks lead this way and that, scattering in all directions—but soon enough you're back to full speed, as if auditioning for a car commercial, blasting across flat desert playa.

Where to stop for the night? My friend and I agree to drive until 8:30 and then stop. Nothing marks the land to differentiate this spot from any other; it's simply chance. The playa spreads all around, taupe scales of clay as far as you can see, an enormous jigsaw puzzle. Deep blue night rises over the range in the east like a billowing sandstorm, while a rosy bloom fades in the west. You feel far from anything. We turn off our phones to keep them from searching for service all night. And with no phones we have no clocks. We turn off artificial time and set up our chairs to watch the night arrive in natural time.

Sting will be here next month for Burning Man—the annual August festival with tens of thousands of partiers and a ritual burning of a giant sculpture the final night—but tonight we have miles and miles to ourselves. We feel like astronauts on the moon must have felt, except we're in shorts and we brought along a glow-in-the-dark Frisbee. We also have beer. (I don't think they did, although I could be wrong. There is the old story about the astronaut Buzz Aldrin, who, after the Eagle had landed but before he and Neil Armstrong took their giant leaps, took out a small communion kit he'd told no one he was bringing along.)

My chair facing west, I watch the earth turn from the sun back into space as though falling back into waiting arms. The daylight's fade reveals the first stars, and suddenly there's that every-once-in-a-while realization: The stars are above us all the time, all day long, but we only get to see them at night. In the Isaac Asimov story "Nightfall," six suns circle the world so that it never experiences night, and people panic when an eclipse of those suns promises darkness. Apocalypse! But here, that promise is welcome. We sit back and wonder what will be revealed.

Meteors! That's what. Astronomers talk about being able to see the Milky Way as the mark of a good dark sky, but for me it's the presence of shooting stars, their sudden ephemeral scratch of light. When a giant shooting star with a long smoky tail burrows its way across our sky, our conversation ceases and we laugh. Two friends, looking out into the universe.

No cars, no engines, no wind or birds or rushing water, no television or radio. It's quiet here, that eternal kind, like you've gone back in time. Train tracks run along the southern edge of the desert, and there sometimes are trains, long snakes with heads glowing yellow bright. But they're so far away that they pass in silence, like a night mirage

After midnight in the Black Rock Desert, after every last ounce of sunlight is gone, we walk in the dark, my friend toward the Big Dipper, I toward the Milky Way. Both come down as though touching the ground, as though just over there, as though if we just keep walking we will have stories to share. Straight overhead, the Summer Triangle shines in three dimensions, and you feel as though you're walking not under but among stars, the night so dark that it's no longer dark, your adapted eyes guided by the faint glow of mud lit by the stars.

We build a fire. Warmth and heat and firelight orange. Suddenly the playa feels like a flat lid, a surface that might implode. The fire burns as though the earth has opened to release its own internal flames. We look at the fire, then at the stars, all of them fires, all around. A thousand trillion fires, as though we have lit our own star here. My pants legs hot, my eyelids and eyebrows hot, too. No idea of time. No moon. We stare at the fire for a while and then remember the stars, stepping away from one to know the other. Later, in sleeping bags set yards apart, we lie awake by the light of embers, with the Pleiades rising for the first time this year in the east, until the red edge of dawn, and our own yellow star.

We live at a time when a place as dark as the Black Rock Desert still exists. But within decades this darkness will exist no longer, unless the spread of light pollution can be stopped. Already there are people working tirelessly—oftentimes volunteering their time—with this goal in mind. Here are five of those people, in five different parts of the world, doing varied and vital work on behalf of darkness.

On the World Atlas of the Artificial Night Sky Brightness, the Black Rock Desert is shown in black—the darkest category—but the light from Reno and surrounding communities has almost reached its edge. And, truthfully, it may already have, as the atlas was built from data fifteen years old. There is maybe no more dramatic portrayal of the spread of light pollution in our world today than these Italian maps of the world at night. In their color-coded portrayal of how light shines outward, like circles of water moving out from a stone's splash, these maps show clearly how most of us in western Europe and North America no longer experience anything close to real darkness. These maps are the reason I have traveled on an old train, its windows open to the spring countryside flowing by, to the small city of Mantua (Mantova) in Italy's region of Lombardy. Surrounded on three sides by lakes, Mantua was one of Europe's greatest Renaissance courts and features a town center in which three different piazzas join together. In 2008 it became a UNESCO World Heritage Site in recognition of its architecture. It is also high school science teacher Fabio Falchi's hometown.

When I first see Falchi he is smiling and waving a baseball cap over his head while he talks on his cell phone next to the Volvo he's parked outside the station. In his early forties, Falchi has close-cut black hair graying at the temples. Trim and dressed in pressed shirt and slacks, he is polite and good-natured. And although he is shy about his English—which he says he learned by reading *Sky & Telescope* as a teenager (and "a little bit in school, yes")—it is excellent. As we walk through this town he loves, he tells me why he gives all his free time to protect darkness.

"I became involved because when I was five my parents gave me a present of a small toy telescope and I saw the moon, and I liked it. When I was in the eighth grade, middle school, they bought me another telescope. Then started the passion for astronomy, and the passion against the enemy of astronomy, light pollution."

We walk at the edge of the old city, across the highway from two of Mantua's three lakes.

"I remember in 1988 I wrote to the bestselling Italian astronomy magazine asking if they could start to take signatures from their readers to ask Parliament to take action against light pollution, and they answered me it was a lost battle. But twenty-five years later we have fifteen-year laws against light pollution."

Much of that success has to do with the organization of which Falchi is president, CieloBuio (DarkSky), and the success it has had in Lombardy. Home to nearly ten million people and the city of Milan, Lombardy is, Falchi says, "like a small nation." It is where much of the Italian gross national product is generated; on its own, it would be the world's seventeenth largest economy.

"Here in this province, the growth of light pollution is stopped," he says. "We have the same sky now as it was thirteen years ago. So it is a huge improvement over the past, where we had a doubling almost every decade." Significantly, the growth of light pollution hasn't slowed because no new buildings have been constructed in Lombardy over the past thirteen years, or no new lights put up. In fact, because of continual increases in power and efficiency, the region's lamps actually now give twice as much light flux as just a decade ago. But because of CieloBuio and groups like it, most of that light shines downward. "Without action," Falchi says, "we would have a sky that is double the brightness that it was ten years ago. So it has been a lot of work for us at CieloBuio, but we have some results."

He knows the work will not end. "Sometimes we are worried about what might happen in the other countries where we haven't the possibility to work and to explain our strategies. We were able to convince our politicians of the need to act and to make laws, and for now it's working. But if the rest of the planet goes another way, it could be difficult to stay an island. We are very few people and it is difficult to work on all the things." He laughs. "In my spare time I have a job. And also I have to work for free on light pollution. My wife is quite understanding, but I cannot work too much time on this without earning anything.

"But it is stronger than me," he says. "I cannot *not* take action."

We are strolling, talking as we go, past an eleventh-century church and into a fourteenth-century piazza. Beneath an arch between two buildings Falchi points to four small black iron rings protruding from the ceiling. A medieval torture device, he says. Four different ropes were tied to a prisoner's four limbs and slung through the rings and then pulled. "Like Guantanamo," he jokes. A few motorbikes zoom past, and people saunter home, savoring the evening air. Church bells ring the hour, and all around us we hear the cries of swallows as they swoop from mud nests under building eaves. "A sign of spring," he says. We make our way slowly, as the evening moves toward dark, toward a favorite restaurant of his on one of the squares.

"I was struck by an article I read in 1981 about light pollution. I was thirteen, and since then I have thought about light pollution and how to combat it. And thirty years later, here we are, still working. And what I think is that if we are not able to solve this small problem—small in respect to other environmental problems—if we are not able to solve this one, well, we will not be able to solve the rest. And nature will solve it for us. It is the human beings that will lose, not the planet.

"In Europe we have arrived at the point where we cannot anymore go easily to a dark place. And if also in the United States they don't take action, or take wrong actions, it will be only a question of time. The growth of light pollution is fast but not fast enough to make people take action. So it is fast, and in one generation you see a lot of difference. But one year to another there is not a lot of difference, and people who are born now are used to this sky and they don't know what they have lost. The elders remember a long time before, when there was a good sky. So it is a strange thing, fast but not fast enough. Or, on the other hand, not slow enough.

"We really don't realize what we're missing. Our children grow without perception of the universe. They grow up without ever having seen the Milky Way, or a pristine sky, a total solar eclipse. And these are places that you need to see, like Venice, like the Grand Canyon. There are some things that open your mind and your heart."

I want to add another experience to the list of things that open your mind and your heart, and that is sitting down for dinner at an outdoor restaurant in an ancient Italian plaza. Falchi points back toward the kitchen, housed behind a brick arch, and smiles. "This building is from the twelfth century, so maybe nine hundred years old." The name is Ristorante Grifone Bianco, and the menu features a Romanesque seal with a white griffon, the mythological creature with the head and wings of an eagle and the body of a lion. The restaurant sits on Piazza Erbe, near an astronomical clock from the fifteenth century, and has a dozen patio tables. At 8:00 p.m. we aren't the first dinner guests, but we do have our choice of a few open tables at the edge of the piazza. Falchi and I both love food, and deciding where to eat has been part of our conversation. Pasta is what I'm after, the local specialties. Beyond that, it's his choice. When we sit, his choice feels perfect, and not only because of the menu.

We shouldn't think that astronomy or even a night sky is the only reason for protecting the night. Pasta and the sharing of wonderful local food with friends in a pleasant setting are good reasons for protecting the night, too. Falchi and I will start with red wine—Morellino from Tuscany and Lambrusco from Mantua—and a plate of thinly sliced meats with sweet onions. And we will continue to talk about Falchi's work. "It is all wins," he says. "If we combat light pollution, you haven't light pollution, you have less energy consumption, you have to spend less money for lighting, for

The spread of light pollution over Europe, circa 1996, as demonstrated by the World Atlas of Artificial Night Sky Brightness. *(P. Cinzano, F. Falchi [University of Padova], C. D. Elvidge [NOAA National Geophysical Data Center, Boulder]. Copyright Royal Astronomical Society. Reproduced from the Monthly Notices of the RAS by permission of Blackwell Science.)*

taxes." Falchi could have mentioned the human health angle. Or the ecological angle. But the point is that the more you learn about light pollution, the more you see that solving it is a win-win situation. Or, as they say in Italian, "it is all wins." And, yes, there are challenges. As Falchi says of his opponents, primarily the energy companies and some fixture manufacturers, "They are not defeated. They try to find other ways to make the things go as usual without any regulations." But the greatest challenge someone like Falchi faces, or CieloBuio, or any of the other advocates for darkness that I will talk with is this: a lack of awareness—awareness of darkness, awareness of light pollution. As it was for Rubin Naiman and David Crawford, so it is in Italy for Fabio Falchi and CieloBuio. It was the same in Paris for an activist there when I asked if the French are more aware than others of the problem of light pollution, living, as they do, in the City of Light. He said, "As anywhere, no more, no less."

And if we are aware, what can we do?

"You can help in a lot of ways," Falchi says. "It depends on how much time and effort you want to put into the issue. If you live already in a place with laws, you can contact your municipality and say, 'There is a road that is not correctly lighted; you have to act.' On our website we have letters already completed. You have only to change the addresses and put your signature, and you are done. In three-quarters of Italy you can do this. And in Veneto, the region to the east of Lombardia, there is a web of amateur astronomers named *Venetostellato*, Starry Veneto, and they make since 2009 about four thousand letters to the municipalities asking for action against some specific installation. And it works. Where you haven't laws you have to work with politicians who will support you.

"You have to be very strong technically, you have to have all the answers to their criticisms. And so you have to know your business. And then you will be successful. You can't go there and only say you have to protect the sky because it's beautiful. It doesn't work."

Falchi chuckles as he says this, and it could be amusement at the thought—what if it were that easy, if one could convince a politician simply by pointing out beauty—but it could also be delight, for the waiter in his black coat, black bow, and white apron has brought to our table the first of our two pastas (we will order four all together, and be too full for *secondi* of meat or fish, though not too full for dessert). The first two are *tortelli di zucca al burro versato* (which directly translates to pumpkin ravioli in "well-versed butter") and *agnoli ripieni*, a small pasta stuffed with beef, pork (salamella and prosciutto crudo), eggs, Parmigiano-Reggiano, bread crumbs, nutmeg, salt, and pepper. These both are specialties of Mantua, though the White Griffon had only the *tortelli* on its menu; Falchi had asked, on behalf of the American, if the kitchen could make the *agnoli*. And here now is their "yes, of course" on the plate before me.

Music from an accordion plays in the background, and I hear from the other tables the murmur of several languages, the muted percussion of silver forks against ceramic plates. A small white dog parades past, barking at something, at nothing, proud to own the moment. We have timed this all perfectly, timed the break in the conversation to coincide with the last of the day's light fading behind the very old buildings. Timed our dinner to accompany the coming of night, and darkness.

But also, unfortunately, our dinner comes with the rising of floodlights on the astronomical clock tower and the piazza, including one that shines directly onto the restaurant's yellow umbrellas. When the waiter returns to check the level of wonder we're experiencing (or whatever he says—I understand only the *grazie* and *prego* and *sí* that Falchi and he exchange), Falchi points over his left shoulder at the floodlight and—I assume—asks about the light. The waiter sighs and launches into an explanation that Falchi replies to with nods and what sound like questions. When the waiter leaves, Falchi smiles. "Until last year there wasn't the flood, but I think I am the first to complain." He laughs. "The waiter said it was too dark last year for dining here. So they think it is a good thing. Maybe somebody else will complain."

Falchi knows the odds are greater that someone will complain that there still is not enough light, whether here in this piazza or just about anywhere else, and that they are more likely to get more light than he is to get more darkness. Part of the problem, he says, is that politicians, while not convinced by beauty alone, can be convinced by the promise of votes alone. "It is proven that politicians are reelected more easily if they make something that is visible," he says, "and what is more visible than changing the lighting? People ask for more light and they give it." Nonetheless, Falchi is trying to get the city of Mantua to turn off the lights on its monuments—like the clock tower—after midnight. I tell him about my experience in Florence, about how I couldn't help thinking that the city could be so much more beautiful at night if it were lit with care rather than with glare. And how I've seen this in too many cities now, including otherwise attractive cities and towns that spend thousands or even millions on improving their daylight appearance but then forget themselves after dark.

During Earth Hour this year the lights on the famous Leaning Tower of Pisa and surrounding Piazza dei Miracoli were turned off, Falchi says. "And there are some photos of the piazza without the lights, and it is fantastic, with the stars. It would be fantastic to have a network of great places like this to be the forerunner of a new way to see the monuments."

Because for most of history that's how you would have seen them, in moonlight or starlight? I ask.

"Or with very warm lights, like a flame," he says. "Not with bright light and white, almost blue, light."

It's a thought I have again and again. How fascinating would it be to see certain old buildings, towers and churches, for example, not plastered with floodlights but touched only by moonlight, by starlight, or even just flames? In almost every case the scene wouldn't be as it had been before electric light because even if—as Falchi saw in Pisa—you turned the lights off on the monument, there would still be the sky glow from lights in the rest of the city or village. Still, that could make the atmosphere more appealing, I imagine—a bit of ambient light to warm the scene.

"I will try here," he says. "If I will be able to shut off the lights at midnight we will see the difference, and maybe it will be a discovery, because no one has seen the monuments with dark or surrounding lights shut off in fifty years."

"People don't know what they're missing," I say.

"But most people, when you explain better, they understand and agree. Not all, but most. We are starting to realize the value of the dark. Up till now we try to exorcise the night. For me it is impossible to think because I love night, of course, but for some people the dark is the dark side of life. They are worried about it."

When our two other pasta dishes arrive—*orzo mantecato* with shrimp and pancetta, and *bigoli carbonara*—I turn my tape recorder off so that we can eat. A man sings now with the accordion.

Earlier, Falchi and I had walked past some new streetlights that might look odd to someone who had never seen something similar before. Although they were in the shape of a regular carriage light, with four rectangular sides, there was no glass—and, um, no lamps. Actually, there were—the Italians are creative, but they haven't yet mastered the lampless streetlight—but the lamps were housed up inside the top of the fixture. The result of no glass and no obvious bulb is significantly less light pollution. These are the fully shielded lights that Falchi and those like him hope will be coming to a lamppost near you very soon.

It used to be thought that if we just kept the light from going straight up, if we could just cap the lights, we could stop sky glow. And while doing this is vital, it's not enough. In the past few years we have found that the worst cause of sky glow is not the light allowed to soar straight up but the light emitted at low angles above the horizontal. Light cast at these low angles is scattered and reflected by the aerosols and water droplets in our atmosphere to a greater degree than light sent straight up. As a result, these rays travel for long distances near the ground and cause sky glow in locations far from the light source—such as in the suburbs and countryside. Light headed straight into the sky causes sky glow near the lights, but not so much far away. Most streetlights are shaped like a bowl or vase, and light within these shapes will bounce every which way, including at these low angles—and into your eyes. While there are plenty of different fully shielded fixture designs, many made—like those in Mantua—to look like those we're used to seeing, one feature they share is that unless you're right up close, looking up at the fixture, you can't see the actual lamp.

I ask Falchi if he thinks light pollution can be controlled enough so that a starry sky can be restored everywhere.

"Well, not really everywhere. But I think that if the technical prescriptions we suggest are enforced everywhere, we can have, even in Europe, places that are a one-hour drive from everywhere where you can see a very good sky. For example, from where I live, if you drive one hour, you arrive in the mountains, and we have a sky that is twice or three times as bright as the natural one, so we have lost the pristine beauty of our skies. But if we work hard to make good enforcement everywhere, then if you go fifty kilometers from major cities you can have a very, very good sky."

This is the exciting part of the question: What can we do? It's not only a question of what actions, but what is possible.

While we eat dessert—*torta sbrisolona*, an almond cookie and Mantua specialty—Falchi tells me he wants to make another map, this one an atlas of what it could look like, what it could be. By using computer simulation of the light that towns and cities would emit if they were using fully shielded fixtures, he could show us the future. It would be a map that would show us what we can achieve. It would be a map of possibility, a map of the world at night as the night could be.

On a Tuesday night I'm in southwestern England, in a town called Wimborne, at the local astronomy club meeting (where for the first time I will hear stars referred to as "chaps," as in "these chaps here in Orion's belt"), and I'm holding a map unlike any I've ever seen. It's a road map, the old-fashioned foldable kind you might once have tramped around with on a trip of the British Isles. Except that this Philip's Dark Skies Map is a road map for astronomers or anyone wanting to find a good place to watch the stars—in other words, it is a map of light pollution and how you might get away from it. Essentially a road map of the British Isles with the appropriate section of Falchi's World Atlas of the Artificial Night Sky Brightness overlaid, it reminds me of activist Pierre Brunet of France's Association Nationale pour la Protection du Ciel et l'Environnement Nocturnes (ANPCEN) telling me in Paris, "I always say that a car is the main observing instrument of the amateur astronomer, not the telescope." That's because the vast majority of astronomers—or anyone else—can no longer see good stars where they live. They have to get in their car, grab their Philip's Dark Skies Map, and hit the road.

I'm in Wimborne at the invitation of Bob Mizon, of the British Astronomical Association's Campaign for Dark Skies, whom I'd first met in London. "The trend is good," he tells me. "More and more councils are thinking about light pollution, and more and more people know about it. I was pleased a couple of years ago when I looked in the dictionary and there was the term 'light pollution.' Twenty years before, nobody would have understood the term."

Because?

"It was something that nobody thought about because everybody had grown up with bad lights and thought it was normal. And when you told them it was a kind of pollution, they would think, well, Why? That's just lighting, isn't it? Surely it's okay. Good stuff. Helps us. Light, good; dark, bad. Says so in the Bible."

Mizon's good nature covers the fact that anyone working to reduce light pollution runs into these same objections—apathy and ignorance, especially—all too often. The term may finally have landed in the dictionary, but acknowledgment of the problem still hasn't reached a critical mass.

"The thing that gives me most hope is the energy crisis," he admits. "We know that energy will become more expensive. As the oil begins to run out and we are forced to use different kinds of energy, people will become a lot more careful about how they use it. Because at the moment, most people don't care. People don't switch lights off, they don't turn the heater down by a degree or two. If people see water being wasted, they will quite often do something about it. But if people walk along a road and see lights shining all over, they don't care. But when your electricity bill comes through the door and it says five hundred pounds, you'll think, Shit, let's do something about this."

It's estimated that the European Union spends some 1.7 billion euros a year on wasted outdoor light. In the United States, the figure is a similar $2.2 billion. Compared to the money we spend for heat or gasoline, these aren't big numbers. But there's no reason to be spending this money—it's all waste, in the form of light pollution. There's also that small matter of cheap—artificially cheap—energy that fails to account for the true cost of its production for human and ecological health. But the bigger point Mizon's making, and an idea that I run into time and time again, is that the way we light our world is going to change.

It's changing already—one of the big concerns for many people is the coming of LEDs ("this tsunami," Falchi called it) before we really understand their strengths and limitations, benefits and perils. But several people told me that the way we light our world in the future will be different from—perhaps radically different from—the way we light it today. I can't think of anyone I talked with who thought this was a bad idea—few who are thinking seriously about lighting and darkness are happy with the current state of affairs. But the question is how we will change. Will we continue along the path we're on, with levels of lighting growing every year, everywhere, and with only a few rare places (such as Lombardy) keeping a thumb in the dike? Or will we somehow begin to use light differently? Will we, perhaps, even begin to value darkness?

Mizon has two stories from his London childhood of society choosing to change course. He remembers the "pea soup" smog of the 1950s ("I remember how awful it was. It must have been like smoking a hundred cigarettes at once, and we thought it was fun!"), and he remembers the River Thames.

"The River Thames is a great success story. When I was a boy the police would come to our school and warn us about the river. Don't go near the River Thames, it's toxic. You mustn't ever try to drink the river. Don't get it on your hands, because if you put your fingers in your mouth you will be poisoned. And it was absolutely true; the river was a sluggish, black filth. And then, there was legislation and people were no longer allowed to pump sewage into the river anymore. Because obviously in Victorian times all the sewage from people's toilets went straight into the River Thames, and it stank terribly. And the legislation was such that the river immediately became much cleaner, and now, fifty years later, there are one hundred twenty different species of fish in the River Thames. It's now a living river, thriving. Just a little bit of legislation causes a big difference. And that's what we say about light pollution. We're not asking for some huge draconian change in the law. Just a little bit of legislation about planning and getting the lights right when you put them in solves the problem."

By "planning" Mizon means building codes. "This will be the solution to light pollution in this country in the long term. Because when you build a new development of any kind—industrial estate, housing development—it has to be approved by the local authority, through the planning system, and there are certain directives which you have to conform to. And all we're saying, quite simply, is that in those directives it should say all exterior light will be directed only on the premises to be illuminated. Job done."

Only on the premises to be illuminated—not into the sky, not onto your neighbor's property, not into the street. It doesn't seem like too much to ask.

"Once you've done that," Mizon explains, "the problem will evolve away over the years. People don't think they're being dictated to, they think they're just obeying a sensible norm. So, get your lighting right, please, and everybody will be happy. It's not the same as saying to somebody, Your light is crap, do something about it."

For his part, Bob Mizon is doing all he can right now, but he's realistic. "It would be nice to wave a magic wand and suddenly get the stars back," he says. "But I realize it's a long-term evolutionary process. We're talking about the night sky in perhaps fifty years' time, when we're dead. But it's still worth doing. Because if my son, who's now nineteen, can see a good night sky over wherever he lives in 2060, then that's what it's about. You have to be patient and a bit altruistic. You're really doing it for future generations, not for yourself."

When the meeting ends, a half dozen amateur astronomers drive up the hill to a local pub for warm beer and crisps (and Mizon's mock disgust at my ordering cold beer). But before going inside we stand in the dark parking lot looking down on Wimborne, population fifteen thousand. Five years ago, Bob Mizon convinced the town council to replace the city's streetlights with fully shielded fixtures, and the effect is startling. Leaving the meeting, walking to the car, driving to the pub, I haven't noticed anything out of the ordinary—we certainly haven't been, to borrow Mizon's phrase, "stumbling through medieval darkness." We've had plenty of light. But now I realize, as we look down to where we've been, I can't tell where exactly the darkness ends and the town begins.

That an entire town like Wimborne—or even an entire city like Paris—could design its lighting to achieve a certain effect owes much to the work of Roger Narboni. When Narboni—who worked closely with François Jousse on the relighting of Notre Dame—formed his company Concepto in March of 1988, the world of lighting design was still in its dark ages. Paris monuments, for example, were often simply plastered with energy-gulping spotlights, with little subtlety or integration into the neighborhood. For one of Concepto's first contracts, Narboni designed for the French city of Montpellier the world's first "lighting master plan," a comprehensive plan for how a city's lights would create beauty and safety together—rather than simply being, as he says, "strictly and blandly functional." Since then, the idea of a lighting master plan—and of the professional lighting designer—has spread around the world. If the future of darkness has everything to do with how we use artificial light at night, then lighting designers like Narboni will have much to say about it, for already they are imagining the future of light.

Born in Algeria, Narboni moved to France in 1962 and lived in the barrio for twenty-five years. When he became a lighting designer he swore he wouldn't forget where he'd come from.

"Believe me, we can be as poetic there as we can in the center," he explains. "So I always try to work for parts of the cities that are the most destroyed, the most difficult. It's easier to bring beauty to this part with light because you can hide a lot of things and have with light a total metamorphosis." Narboni says lighting the tougher parts of a city is often more rewarding than lighting landmarks like a cathedral. "A cathedral is already beautiful," he says. "It doesn't need any lighting. And also, the way people receive it is totally different. When you work on a historical center, everyone says, 'Oh, again,' 'Whatever,' 'This isn't terrible'—people are spoiled." But, he says, whereas people in the wealthier areas tend to take his lighting design for granted, when he works in poorer districts, the citizens of the neighborhood are very appreciative. "They are like, Wow, our district could be like that? And they thank you, they kiss you, they react totally differently. It's like you changed their life." Narboni says he recently went to check on one such project, a simple decorative lighting scheme projected against the side of a building, and met a local man coming out of a bar. "He was looking at the lighting and I asked him what he thought of it. And he was totally drunk," Narboni says, laughing, "but he answered, 'I live here, and I like to look at that because it's so poetic, and we deserve poetry, too.'"

While he has lived in France for forty-eight years, Narboni says his Algerian culture continues to have a powerful influence on his designs. "In North Africa, shadows are more important than light—the way that you play with the contrasts of shadows and light—because it's hot country, we protect ourselves from the sun, we never go to the beach to get sun, you know. We hide ourselves because the sun is so powerful. So for me in my everyday work this play between shadows and light is very important."

Unfortunately, he says, people are afraid of shadows and darkness. His dream is that we could have an educational program about light and darkness in the schools, even in kindergarten, "because the kids don't learn anything about light from their educational program—they learn how to play flute, they learn how to do crazy things—but light, no one is talking about it. About darkness the only thing they get is what they read in the tales, and very often it's related to the devil, to fear, and it's a pity, because they don't learn how to play with shadows, how to be peaceful in darkness."

He soon will have a wonderful opportunity to help people become more comfortable with darkness, as he recently won a competition to design the new lighting master plan for the city of Paris.

"There is a new policy in Paris that calls for renewing all the urban lighting while reducing the energy consumption thirty percent by 2020," he explains. "The idea is to have a new look at what should we keep, what should we stop, and what should we create for new lighting. So it's tough, because they want new things, beautiful things—you know, City of Light and everything—and then the energy consumption should be very low." Narboni says he used studies done by the urban planners of Paris showing the hour-by-hour occupation of the city during the night to do his redesign of the urban lighting. First, he proposed shutting down some of the architectural lighting (Paris has more than three hundred elements to light, such as buildings, fountains, statues, trees, and 32 lighted bridges, plus all the street lighting), and then he got creative.

"The main idea we propose is to not have an even level of light everywhere, and this is a revolution because in Paris every single street is even in terms of lux level, whether it's a small one or the Champs-Élysées. So the idea is to ask, Why should we stay like that? The second idea is to dim, depending if there are cars or pedestrians or not—if there is no one in the street, why should we light the street? And for that we are making a lot of studies about where is the nightlife in Paris—the night geography of Paris. The idea is to understand better the morphology of Paris at night and then to dim the lux level depending on the level of activity. Another idea is to put on the lighting ten minutes later every single day, so to get Parisians used to having a little darkness, and multiplied by three hundred sixty-five days, that's a lot of energy consumption. And again in the morning we will cut five minutes earlier, maybe ten—we'll try."

The future of light can be very interactive, Narboni says. "I'm sure in ten or twenty years everything will be automatic. And if you are there, there is some light, and if you are not there it goes down and no one cares. If we can have the right light in the right place for the right person it might be a very nice future. So let's dream."

Dreaming of new ways to light the night is something Nancy Clanton has no trouble doing. As the founder and president of Clanton and Associates, a lighting design firm in Boulder, Colorado, that emphasizes sustainable design to save energy and conserve darkness, her excitement for the future of lighting is palpable. Like Roger Narboni, she envisions a future full of interactive lighting, a world where the growth of the electric car industry has a direct effect on our use of lights at night. She seems unfailingly optimistic about a future where we change our common assumptions about lighting.

Take light posts, for example. "We should really evaluate whether lights on poles are the best way to light a city or an area," she says. "For one thing, light poles are expensive, and people hit them. Seriously, you put a pole up, and someone will hit it. Communities are anxious to get rid of their lighting poles." The reason? Money. When you break down what it costs communities to provide street lighting, more than half the expense comes from buying and maintaining the poles. Instead, Clanton envisions relying on different layers of light—headlights, step-lights at foot level, motion sensors near crosswalks. "And," she says, "layers may be added or subtracted as the night goes on. Maybe the overhead lighting is turned off and just the low-level sidewalk step-lighting is left on in the middle of the night. I would really like more options in lighting instead of just one size fits all."

Much of her optimism about lighting comes from what she sees as the possibilities presented by LEDs. Unlike current streetlights, which are either on or off, and at one level of brightness, LEDs can be set for different levels of brightness depending on the situation. Earlier in the night they might be set brighter; later in the night, they might be dimmed. Combine that ability with the potential of smart grids and computer controls, and communities could have the ability to have different levels of light for different areas. "A neighborhood could say, 'Hey! We really don't want lights on anymore after midnight,'" she tells me. "So someone could go to Google Maps and group all those lights in that neighborhood and say, 'Okay, go down to ten percent or five percent.'"

What might drive a community to ask for lower lighting levels? Again, money. Whether you live in Boulder or Paris or Wimborne or almost anywhere else in western Europe or North America, peak energy use comes during the day—air conditioning on a hot afternoon, lights and heaters in winter. But energy companies have already paid for their generators and other equipment and would like to have them running at a high level twenty-four hours a day. At night, when energy use drops, energy companies have no way to replace that demand, so they encourage our use of electric lights. In fact, Clanton sees this as "the untold story" about how much outdoor lighting we use—it doesn't have nearly as much to do with safety and security as it does with the need energy companies have to even out their peak load.

Pete Strasser, technical director for the International Dark-Sky Association, agrees. "Utilities pretty much need to have streetlights at night. That's a load. That need to keep minimum spin going on those generators to ramp up in the morning. They can't shut them down real low. They have to keep going. So what's been the solution for a hundred years? Streetlights." Clanton, Strasser, and others told me that once electric cars come fully online, cities will take a second look at leaving the lights on all night because the cost of electricity will so dramatically increase. "Power sold at night, it's cheap as hell," Strasser explains. "One or two cents a kilowatt hour, wholesale rates, because they're just dumping it. I guarantee you it's not going to be one and two cents a kilowatt hour when you're plugging your car in at night. That's going to go heavy retail price—ten, fifteen cents an hour. So, as soon as there's a tipping point with electric cars, you're going to hear, 'You know what? Studies have shown that those streetlights really haven't been necessary. Studies have shown that crime rates have done nothing with streetlights.' And it's coming. Because you're right—they haven't."

In the meantime, Clanton is keen to help show people that we are using way more light than we need to. One of her prime clients in recent years has been the U.S. military. She is helping it understand that instead of high levels of light, contrast is the most important issue when it comes to visibility—and therefore security. "We did this experiment with the antiterrorism group with the navy. They go to bases and try to break into the buildings, evaluating defenses. And all their high-security buildings are really brightly lit—walls, the concrete—everything was lighted so the cameras would work. So, if you are going to break into a building, what color uniforms would you wear? They were all wearing black. I told the squad commander that he and I should wear white. Guess what? No one could see us because we were blending into the white concrete and the white building. Anyone wearing black was a huge contrast.

"You want to see objects in varying light levels or varying

backgrounds, kind of like what prisoners used to wear—black-and-white stripes—so no matter where they were, you could see them. In a dark situation, you see the white, and in a bright situation, you see the dark. That is why the best thing you can do if you are jogging at night is wear something really bright and something really dark, so you can always be seen."

Talking with Clanton I am reminded of the small Wisconsin town where I lived, and how Luna and I would go for a walk in our neighborhood every night at 11:00 p.m. before bed. We'd step down to the sidewalk and take a right, going three blocks to where the town's middle school stood blazing away in the Northwoods night. At regular intervals around the brick schoolhouse building, high-pressure sodium wallpacks were glaring straight into a viewer's eyes, spraying light across the streets, completely illuminating the facing neighborhood houses. As Luna and I followed the sidewalk around the school, we were steadily bathed in light. It was the best example of light trespass I have ever seen, and a perfect example of 1) decisions about lighting made during daylight, installed during daylight; 2) the people who lived around the school having to accept—if they noticed, and maybe they didn't—the light plastered against their house sidings; and 3) a needless waste of money and energy that everyone in town paid for in taxes and the loss of their night sky. This small town could have wonderful dark skies if this kind of wasted light were controlled. In fact, small towns all across America could save money and bring back their stars by paying attention to schools like this (not to mention warehouses and businesses), lit without thought and then forgotten.

"Wouldn't it be great," Clanton says, "if in a schoolyard at a certain time in the night, all of their lights went off and switched over to motion detectors?" She tells me that the Loveland schools north of Boulder had tried this and it had worked out well. "The police loved it," she says, "because if they saw the property with no lights on, they knew no one was on site."

The best exterior lighting, whether for schools or most other kinds of buildings, Clanton tells me, would be "responsive to your actions" in the way she'd described. In this way, a building would be dark unless there was someone there, and then, anyone in the area would be more likely to notice the light coming on and take a moment to look. As I walked around the Wisconsin middle school, I used to think, *What if this community chose instead to light this school with care?*

"You know," Clanton says, "increasingly, the higher-end the community, the more subtle the lighting. If you go to Aspen or Vail, everything is very subtle. And people go, 'Safety and security. We need more light.' But the people in Aspen and Vail have more assets than anyone else. So, I am not buying that you need more light for higher security and safety. I think a higher lit area signifies a less desirable area. Like with retail or hospitality, you go to a really cheap hotel or restaurant like fast food, and it is brightly lit with a lot of glare. The finer the place or the higher-end, the more subtle everything is."

While talking with Clanton, Narboni, and other designers, the word I often think of is "progress." I wonder if, as we ponder where to go next with lighting, we might think of progress in a new way, as defined by subtlety. Bright lights have long been seen as progress rather than as light pollution, but that can change. To prefer the stars doesn't mean to relegate yourself to living in the Stone Age at night. Good lighting is effective lighting is subtle lighting, and good lighting prefers the stars.

Of all the places from which to see the stars, a city parking lot is not high on the list. But as I step from the Lowell Observatory in Flagstaff, Arizona, there is the Milky Way bending over the mandarin orange of low-pressure sodium lights. You might expect the lights around an observatory to be well shielded and subtle. But that Lowell lies within the city boundary just up the hill from downtown is dramatic testimony to the city's strict lighting code, arguably the best in North America, maybe in the world. From the observatory's site, the city of sixty-five thousand spreads toward the eastern horizon, looking noticeably darker than one would imagine a city of its size would look. "It's not perfect," says my host, Chris Luginbuhl, "but it's proof of what can be done."

No one has taught me more about lighting, light pollution, and the importance of darkness than Chris Luginbuhl. An astronomer at the U.S. Naval Observatory in Flagstaff, Chris introduced himself after I'd spoken at the IDA conference several years ago, where I'd introduced—to a room full of engineers, astronomers, and lighting industry folks—a book of creative essays about the value of darkness. At the meeting, I'd read from Henry Beston's *The Outermost House* and then asked if anyone knew what year Beston had written of "lights and ever more lights." Chris raised his hand and said 1928. Since then, each time I've been to Flagstaff I've checked in with Chris and his work on behalf of dark skies.

While the tradition of protecting dark skies in Flagstaff goes back officially to 1958, much of its success as the world's first "dark sky city" is thanks to Chris. In his role at the observatory answering questions from people all over the world about lighting and ordinances, his advocacy work with the IDA, his volunteering locally—speaking to city councils in Flagstaff and other towns—and his scientific papers (he was one of the first to identify the need to fully cut off the shallow angles of light from fixtures), Chris has done enormous work. The fact that he's both a scientist and someone who can quote Beston—or Joseph Wood Krutch, John C. Van Dyke, or Rachel Carson—speaks to his well-rounded approach.

And so it is hard to hear him speak when we get together this time. For Chris has always—at least in the years I have known him—been optimistic about the challenge of controlling light pollution. In short, in our many conversations he has convinced me that light pollution is a problem that we can solve.

I still believe that—and I think he does, too. But when we get together this time, he speaks about "my former optimism," and I hear in his voice a doubt I haven't before.

"Night is beautiful as it is," he says. "We need to understand when we're making trade-offs. There's beauty in artificial light, I'm not denying that. But it's a problem to assume that if you do quality lighting, then it's compatible with preserving the beauty of the night. It's not. You can light a town absolutely beautifully and lose the sky."

Astronomer that he is, Chris has never been concerned only about losing the sky. As pleased as he is with how the citizens of Flagstaff embrace dark skies as part of the city's identity, he's often frustrated that many people think the lighting ordinances are only for the benefit of the astronomers. "It shows how far we have to go," he says. "It's like asking why the Grand Canyon is important and saying, 'Oh, well, we need to have that so that the geologists can study the rocks.'"

We stand in the parking lot of the restaurant where we have eaten dinner, a couple blocks from the downtown center. Lit to code with just three low-pressure sodium fixtures, the lot is bright enough that we can see just fine, though to many Americans it would probably seem dim.

"Just doing 'quality lighting' isn't going to work," he continues. "You can light the whole world according to lighting professional standards and we will lose our skies. And I don't think that's enough. It doesn't address the more fundamental problem of when do we need lighting and when do we not need lighting."

I have heard this argument from Chris and others before: If we perceive our choice to be only between "good lighting" and "bad lighting," we forget that "no lighting" could also be a possibility. In fact, in plenty of cases, no lighting would be the best choice among the three alternatives, but seldom are we offered that choice or remember it's ours.

This has enormous ramifications when you consider that, everywhere you look, human populations continue to grow. It's almost a given that with more people there will be more lights. And even if all the new lights are fully shielded, they still are new lights. Recently, Chris published a study in which he plotted a number of towns and cities in the Southwest according to population size and lighting level. Not surprisingly, the chart showed a steadily upward trending diagonal line: as towns grew in size, their level of light grew as well. The two exceptions? Flagstaff was some 25 percent below the line, and Las Vegas was way, way above: Its level of light was actually double that of any other city its size. Because of its strict lighting ordinance, while Flagstaff's population grew by almost 25 percent from the years 2000 to 2010, its level of light increased only about 17 percent. Chris estimates that, because so much of the new construction during that time was heavy lighting users such as motels and restaurants, Flagstaff's level of light might have grown as much as 40 percent to 50 percent without the lighting ordinance. What frustrates him is that even with the country's strictest lighting codes and a tradition of consciousness about dark skies, Flagstaff has still grown brighter.

"Maybe what we've settled for is that tomorrow it will not be as bad as it would have been, but it will still overall be worse," Chris explains. "Some people might be satisfied with that. I'm not. I'm discouraged by it. To say that we worked as hard as hell and we're 17 percent worse rather than 40 percent worse—it's hard to be motivated by slowed loss [of darkness]. It's progress, but not in the sense that the sky's getting darker, unfortunately."

In fact, when I ask if he thinks every place is getting brighter, he says, "That's my impression. As a scientist I don't want to say I'm certain when I'm not, but I would challenge the world of dark sky activists to document a place that has gotten better. I think that's a fair challenge. It's difficult to do, and I think it may not be doable." Chris worries that while communities may have small successes here and there with controlling light, the overall battle is being lost. "You could say, 'Oh, we've got a nice law in place, and the shopping center that was built under that law is, oh, so much better than it would have been.' But it's still lighted. I'm asking for a sky that's gotten darker."

As Chris and I drive east out of Flagstaff toward his home behind Humphreys Peak, we pass bright motels. ("'We'll leave the light on for you,'" Chris reflects. "It's a nice sentiment until you think about it.") We see wallpacks that must have come from the same factory as those I endured in northern Wisconsin. But we also see things that you would not see in most other American cities, including major thoroughfares without streetlights, backlit local business signs turned off after-hours, and gas stations lit not to daylight but simply to the level you need to pump gas and clean your windshield.

Blazing bright gas station canopies—Chris likes to joke you could perform surgery under their lights—have become so ordinary that those lit to Flagstaff's code seem, in Chris's words, "startlingly faint." Feeling adventurous, I ask him to pull under one such canopy to see if I can actually fill his gas tank in such dim light. "From twenty years ago till now," he says, easing next to a pump, "the level of light at most American gas stations has gone up by a factor of ten. Here we have a standard that is much lower than that, but quite reasonable." I can vouch for that—I have no trouble reading the pump or filling our tank. While the canopies at most stations do block light from escaping skyward, the lights under the canopies are like any others: Unless recessed into the canopy, their light will shine at low angles that add unnecessarily to sky glow, trespass on neighboring properties, and cast glare.

But while gas stations are bad, parking lots are worse. As with gas stations, shopping center and business plaza parking lots are often lit ten times as brightly as they were twenty years ago. But because of their larger size and number of lights, parking lots are often the brightest sources of light in any community. Here in Flagstaff, that's not the case. Even the big box stores such as Target and Walmart conform to the city's standards. "Now on the left here, that's a new shopping center and it's all great," Chris says, pointing. "Although I do see some lights there that are unshielded, but low output." And more surprising—we pass shopping center parking lots with their lights simply turned off. As we drive past the row of car dealerships so common on the edge of American communities, I'm reminded of the Ford dealership on the way into that small Wisconsin town, and how I knew I was close when I saw the bloom of white light on the horizon miles ahead. Here in Flagstaff, we drive past lots where the tallest lights overlooking the rows of cars and trucks have been turned off, the lot lit only by lower-level lights a dozen feet high.

Because parking lots produce more than 50 percent of the outdoor lighting we use, controlling their use—making sure they are shielded during business hours, then turning them down or turning them off after-hours—can have significant impact on a community's light pollution. Unfortunately, in so many towns and cities, our parking lots are lit by wallpacks hung on buildings or by tilted floodlights shining every which way. And this is the middle of the night, and these lots are empty. In the United States, no matter where you live, you almost certainly live within close range of such a lot. But you may not have noticed—these poorly lit empty lots are ubiquitous, so we often no longer do.

That is, until you start to see them, and then you see them everywhere.

Which actually makes me optimistic. I mean, here is a good-sized American city showing any other city it can be done. An American city with a lighting ordinance that (mostly) works—and the streets aren't filled with criminals and the economy is doing relatively fine. The way we light our streets, parking lots, gas stations, schools, cities, and towns is up to us—and Flagstaff shows one possible choice. The lighting here is far from perfect, but already it is better than in every other American city.

Which leads to the question: What's really possible? Can we really bring back dark skies? I remember coming back from Canada into the United States on the way home from Mont-Mégantic, and the customs officer asking me about my work. When I talked about efforts to control light pollution, he smirked and asked, "What are you going to do about cities?" It's a good question, right? And, if you're thinking about the possibilities for controlling light pollution, it is one you have to consider. When I ask Chris, he immediately makes a distinction between cities and towns, saying that in a large metropolitan area there are still plenty of good reasons to do light right, but bringing the stars back isn't at the top of the list. "You may change it from one or two dozen stars visible to three dozen or even four dozen, but all of those are still horrible skies. You don't get the impression of the universe, the sense of place, the sense of scale."

That said, because of scale, the benefits from the other reasons for having good lighting are even greater in big communities. "If we talk about saving energy," Chris explains, "by not wasting fifty percent of your light, how many dollars do you save? Flagstaff is sixty-five thousand people, but Chicago is a hundred times bigger. Numbers like that start to get people's attention." The same thing is true with issues like interference with people's sleep and other health concerns—those stand to carry more weight in large cities simply because they're of greater concern for greater numbers of people.

"The other thing," Chris explains, "is that even though you won't bring the stars back in downtown Chicago—the inspiring, moving experience you can have with a dark sky—if you darken the city, that threshold is going to move closer to the city, and you're going to uncover big suburban areas where lots of people live. Right now, a suburb of Chicago could turn all the lights off in their town and they won't get their sky back. But if Chicago cut its light level, those suburbs could get back their starry sky."

This is the kind of optimistic thought I'm used to from Chris. So, huge cities may never have Van Gogh skies again (no nightly reverie for starwatchers in cities from Las Vegas to London), but they still can benefit in numerous ways from choosing efficient, effective light. And as these cities do—which they have every reason to—the suburbs, towns, rural areas, and wildness around them will benefit, too. This would be the opposite of the stone-in-water effect with light pollution we have now, where the overabundance of city light radiates outward. As the radiating pressure of too much light in the city is reduced, darkness would ripple back inward, returning to the suburbs and countryside some of what has been lost.

We drive north out of Flagstaff on Highway 89 to Wupatki National Monument where Chris knows we will find darkness, and as we walk from the car to a bench overlooking the surrounding desert, I ask about his recent doubt.

"Right now I'm coming to grips with the idea that, despite all the good reasons for doing things better with lighting, it's not happening," he says. "I don't know where it leaves me. My goal has always been to make it better, and I've slowly come to realize that we're not achieving that.

"Once you get aware of what lighting is, it doesn't take much education to realize how carelessly it's used. All you see is bad lighting. And it can wreck your life. You go out at night and all you see is bad stuff. And I don't want to live like that. But I don't know how else to do it."

Chris and I have talked enough over the years for me to guess that he isn't just despairing about light pollution. Always a fan of literature—he once sent me a file of fifty quotes related to his work on behalf of the night—Chris asks if I know John C. Van Dyke's *The Desert* from 1901. "To speak about sparing anything because it is beautiful is to waste one's breath," he begins (reminding me of Falchi saying it wasn't enough to simply tell politicians something was beautiful), then paraphrases the rest of Van Dyke's words. I've chosen half the passage Chris quotes from Van Dyke as the epigraph for this chapter. Here is the other half:

> The "practical men," who seem forever on the throne, know very well that beauty is only meant for lovers and young persons—stuff to suckle fools withal. The main affair of life is to get the dollar, and if there is any money in cutting the throat of Beauty, why, by all means, cut her throat. That is what the "practical men" have been doing since the world began.

"I realized one morning when I got in the car the litany of things that I see and think about as I drive out to the observatory," Chris says. "I see the forest fires, I see the hills across the highway where the off-road vehicles are tearing up the landscape, I see the haze in the air, I see the power lines that are so ugly, I see the areas that are being developed—and I don't see the beauty anymore. One thing after another, I see the wounds. And I don't want to live like that. But what do you do—do you die? Or do you find a way to stick your head in the sand a little bit? Or do you find a way to find beauty?

"I don't know," he says. "There is still beauty."

In silence we watch as, from behind and above the silhouettes of ponderosa pines swaying against the darkest blue-black sky, emerge dozens, then hundreds, then thousands of stars.

1

The Darkest Places

This is the most beautiful place on earth. There are many such places. Every man, every woman, carries in heart and mind the image of the ideal place, the right place, the one true home, known or unknown, actual or visionary.

– Edward Abbey (1968)

"We're going to be in a big black hole on the map," says Dan Duriscoe, slowing his red Toyota Tundra so we can gaze down into Death Valley National Park. "There's nothing between us and that mountain, and then there's nothing for another hundred and fifty miles." Duriscoe, who speaks in a low, gravelly voice and has a penchant for colorful language, has encyclopedic knowledge of the desert West. He's full of dirt roads leading off into deserted valleys, and turnoffs no one else knows. As a founding member of the National Park Service's Night Sky Team, Duriscoe has traveled all over the United States documenting levels of darkness. Death Valley holds some of the darkest places he's seen. "I've been to probably two hundred different locations in the parks doing measurements," he tells me, "and there's only three I gave Class One. One of them was here." Tonight, we are headed to one of his favorite spots, Eureka Valley, between the Last Chance Range and the Sylvania Mountains. "This time of year, fuck, there's not going to be anybody out here," he says. "This is about as isolated as you can get in California."

We drop into Eureka Valley, hit washboard gravel for miles, turn and climb a hundred yards past a Road Closed sign, and park. Immediate stillness and quiet, no wind, no bugs, the scent of sagebrush and creosote, distant sand dunes six hundred feet high. We set up chairs and a table and make a fire. "This is what I live for," he says. "I can't imagine life without this." In the west, Venus is a brilliant white ball just above desert mountain silhouette, bright like a porch light or like a headlight coming over the ridge. But there is no house, and there is no car. A loose chain of flights bob toward San Francisco far to the north, a faint amber glows from Los Angeles southwest, but there is no one anywhere for miles around, and not a single individual artificial light in any direction. Already the sky feels ancient—big, darker each minute, and filling with light, as though the growing dark is sifting stars, spreading them on black fabric before us.

Primitive darkness. The desert before civilization, before settlement. The dark land with no light of its own, and stars coming all the way to the ground: the Big Dipper setting, revolving into the northern horizon, Orion rising from the southeast with Betelgeuse flashing its red-orange cape in the atmosphere. The zodiacal band, like a fainter Milky Way, twirls skyward from the western horizon. The valley so dark you see by night's natural light—the zodiacal light and airglow, and maybe 10 percent from the stars. Duriscoe and I see each other faintly. With no trees or woods we see in all directions to where mountains saw jagged horizons from the bottom of the sky. That sky becomes brighter and darker the longer we stay out, in a way almost no one in America experiences now. Our eyes go dark-adapted, good at ten minutes, even more so at forty-five; but then, after two hours of wide-open eyes and the land with no lights, the sky shifts into focus, like an optometrist switching a lens and saying, "Better?" Before this there were stars, but now there are stars upon stars and a sense of stars you can't yet see. "In the city you will never see like this," Duriscoe says. "Even out here it takes patience, and we expect instant results. People drive out from Las Vegas because they hear there's a star party and say, 'Now show me the night sky. I got about five minutes.' "

We climb from Eureka and drop on winding washboard to Crankshaft Junction. "These are the last of the wild roads," Duriscoe says, "and we're in the blackest part of the map." We are within a few miles of the Nevada border, the Los Angeles light dome blocked now by mountains, but not the faint dome from Las Vegas some 160 miles southeast. I have been at the center of that, and now I am at the center of this—from the brightest spot on the map to one of the darkest.

This dark land. The way I no longer expect to see any lights. The way the dark feels both comfortable and comforting as the night goes on. "Lots of amateur astronomers don't care, as long as they can see the sky," Duriscoe explains, "but to me, it's the land and the sky together that makes this experience unique in the West—the wild land and the wild sky." And then, speaking of the Night Sky Team, he adds, "That's what we're trying to preserve, the ability to see and appreciate the natural night landscape."

At the crest between valleys I take out binoculars, and in turn they take my breath, multiplying the number of visible stars tenfold. I feel as though I'm falling, and I have to pull away to find my balance in the dark. The ground on which I'm standing, the cloth of stars above. The great nebula in Orion's belt, the Pleiades, Jupiter so bright and clear it makes me laugh. And then here comes Sirius. The brightest star we ever see, and—because it's so low, the atmosphere a prism—flashing like a pinwheel sparkler, green and red and purple and blue. Then super-bright shooting stars, like green-yellow flares falling from the sky. And then for me, for the first time, the Andromeda Galaxy in clear detail—the most distant object we can see without a lens, at two million light-years away—the photons that have been traveling toward Earth all this time now touching the back of my eyes. For Duriscoe, it's this kind of firsthand experience with the night sky that matters, rather than seeing it, as he says, on "these goddamn computers, fuck that, that's just totally impersonal hogwash."

He has spent several hundred nights outside. What are some of the best night skies he's seen? "Mauna Kea was thrilling because of the overpowering brightness of the universe; it's just raining down on you. Or Big Bend, where there's not a trace of a light dome. Or after the winter storms in Sequoia National Park, I remember being up at seven thousand feet after two feet of snow and coming out at eleven p.m. with the stars just razor-sharp. Not a drop of moisture in the air. Ten below. Spectacular." People are always asking him—as I have—where's the best sky; where's the best place to go to see a truly amazing starry night? "The best night sky," he says, "is the place where you were when that happened. I can't tell you where to go on any given night, at any given time of the year. I can tell you where to go where your probability of it happening is good, but that doesn't mean it's going to happen. But so be it. That's life."

The turning Earth, the presented universe—in the dry desert air the stars come down to the horizon, in the west blinking out as they fall from the world's edge, and in the east blinking on, as though lit and set into the sky by some happy wild creatures just on the mountain's other side.

When the first national parks were created in the late nineteenth century, conserving and protecting darkness wasn't on anyone's mind. It would be decades before electric lights were used in most of the parks, and even longer before the light from American cities

The Milky Way bends over the "Racetrack" section of Death Valley National Park. *(Dan Duriscoe, NPS Night Sky Program)*

and suburbs began to seriously concern astronomers, scientists, and everyday stargazers.

Times have definitely changed. The lights from Las Vegas, for example, are visible on the horizons of at least eight national parks—and the National Park Service now includes darkness as one of the resources it is sworn to protect. In 2001, the NPS created its Night Sky Team to help raise park service employees' and park visitors' awareness of the importance of darkness, to measure the level of darkness in the parks, and to gauge how rapidly the resource was being lost.

If not for Chad Moore, there might not be a Night Sky Team. I first met Moore at the IDA conference in Tucson five years ago. He was based then in southern Utah's Bryce Canyon National Park, working closely with Dan Duriscoe and two other National Park Service employees, Angie Richman and Kevin Poe. In 2009, the NPS created a new science division combining programs in natural sounds and night skies, and moved Moore to Fort Collins, Colorado. The Night Sky Team now has six full-time members and has documented the night sky in 88 NPS properties, with a target of 110. With more and more national parks adding their services and other parks needing to be rechecked, the Night Sky Team will have its hands full for years to come. And it all started with Moore realizing that the night sky over the park where he worked was growing brighter.

"In 1999, I was working at Pinnacles National Monument in central California," he tells me, "and it seemed that in just the three years I'd been there the sky was getting brighter, particularly in one direction." A new prison had been built near the park, plush with floodlights and wallpacks, and a new housing development had gone in as well. "I thought, Wow, if I can see this much change in three years, imagine what it will be like in thirty years." Moore wondered if there were a way to measure the light pollution around the parks, and he began asking colleagues at other parks if they knew. "I literally got the same answer from a dozen different people: I don't know, but I'm worried about it too. So, I thought, I'm going to figure out how to do this. Maybe it's just my job." Moore wrote a grant proposal to purchase equipment and contacted Duriscoe to see if he wanted to join in. The two of them decided to buy a CCD camera (a research-grade digital camera) and take pictures of the night sky over different parks, figuring it would take a few months to gather the data they would need. That was twelve years ago. Moore laughs. "Neither of us had any idea how much work it was going to be."

Taking pictures of the night sky over the parks isn't as easy as it may sound. For one thing, it's something that can only be done during the dark nights around a new moon, when the moon's monthly cycle keeps its light out of the sky. Next, the weather has to cooperate—meaning no clouds, and preferably no heavy wind; certainly no rain, thunderstorms, tornadoes, snowstorms, or hurricanes. Third, in order for the camera to give an accurate picture of the night sky darkness over each park, Moore and Duriscoe had to get away from any lights associated with the park. There would be no setting up of the camera in the visitors center parking lot and then sauntering back to the warmth of motel comfort. In park after park, the two men carried the heavy equipment out into the darkest location they could find, often spending hours hiking to their remote location, and often spending the entire night outside.

Moore says they were looking for numbers. "Our original purpose was simply to quantify the resource, which was important to understanding acceptable change. As humans we tend to only value things we can give a number to, and one of the tough things about managing darkness is that there's no way to track it. Our contribution is that we have a sophisticated way to accurately do that." At first, he and Duriscoe relied on quantifying the level of darkness wherever they went by using the 9-to-1 rubric of John Bortle's scale. But as they continued their work, they gradually gained a deeper understanding of the different factors that contribute to the quality of a given night sky, eventually developing a new "sky quality index" using a 1–100 scale. Still, their ultimate goal remains the same: to accurately gauge the quality of night sky darkness in order to help the NPS value and protect its resource.

The challenge of measuring the quality—and thereby the value—of darkness remains. "It's not like measuring arsenic in drinking water," says Moore, "where you know there's a certain threshold that you can't cross before it gets dangerous." Even so, having a way to talk about darkness in numerical terms can go far in helping people understand its value. "Humans, we don't detect gradual change well at all," Moore says. By measuring a park's level of darkness repeatedly over time, Moore and the NPS will be able to tell park superintendents, "This is how much you've lost," or "This is how much better the sky has become." By doing so they hope to fight what Moore calls "the problem of us sort of forgetting how good it used to be, where we have this sliding bar of what's an acceptable quality getting lower and lower." It's what psychologist Peter Kahn calls "environmental generational amnesia," a situation where "the problem is that people don't recognize there's a problem" because they don't know any better. In other words, if you have never known a night sky any darker than the one you have now, why would you think anything is wrong?

I am heading to the summit of Cadillac Mountain in Acadia National Park in Maine, first to watch the sun set, then to watch the night rise.

Most cars are coming down; my car is going up. At the top are tourists, but fewer and fewer as the light fades, as though there's some cosmic balance that everywhere must be obliged—the more the darkness, the fewer the humans. There's a parking lot, but no lights—except for those of fireflies, floating around the bushes. I find a good perch and survey the sky—in the west charcoal red, in the north rain-cloud blue, in the east the purple haze of dusk and the same in the south. The orange-pink of high-pressure sodium sprinkled here and there in the islands around the mountain. I know the ocean from the sky only by a darker shade of gray at the horizon. The last chickadee sings as the first stars arrive, and I eat my dinner looking east to open sea, writing by red headlamp and wearing both my jackets, rain and fleece.

By 10:00 p.m. I'm alone on top of this mountain in a national park, lying back on Cadillac rocks. In the southern sky, windows open periodically in the clouds, each time revealing a different constellation. First, Sagittarius, the teapot toward the center of our galaxy—then that's gone, the window closed. Another window opens and Scorpius appears, shining bright, its time on the stage; then, its time gone.

When the rain starts to fall, I start to think that this is not smart, standing exposed on this rock. But for a moment I will, just to feel the rain on my face and legs and arms. Not so far from here, Henry Thoreau climbed Mt. Katahdin and wished for "Contact! Contact!" That desire to know the wild. When the rain first hit, my first thought was to flee, but then I have come a long way to see this night, and not every night has to be clear and starry—at least not for this one-man Night Sky Team. The rain, the wind, the storm clouds moving through, this busy tourist spot has become wild again, this flat rock a theater, only I am the audience and the actors are all around.

The native people here are called the Wabanaki, the People of the Dawnland: Cadillac Mountain is the first location in the United States to be struck by sun each fall and winter morning. Tonight it will be, I suspect, the last place to see the stars, at least around here. To the east the ocean and sky, their black shapes blended, the horizon gone. But to the south, open space remains where bright stars come down to ocean water. I am happy to have my red rain jacket, happy to be otherwise soaked. I stand on this rock in this park, watching until storm clouds move over this last open space, the sky closing down, the land going dark.

In the morning I meet John Muir at the visitors center, his words alive on a sign next to the stairs: "Everyone needs beauty as well as bread," he argues, "places to play in and pray in where nature may heal and cheer and give strength to body and soul." Farther along I find Sigurd Olson far from home, and he and I, two Minnesota boys, stand face to face. "If we can somehow retain places where we can always sense the mystery of the unknown," he says, "our lives will be richer." Beauty and mystery: intangible qualities we all know are valuable but don't always know how to value.

This place gives me hope. On a weekend summer morning a constant stream of people come and go, volunteers directing traffic, rangers offering talks and advice. This place—this national park—gives me hope because of where it lies. Here in the east, not six hours from Boston, a place where for millions of people each year, the beauty and mystery of night can be brought within reach.

"Not a lot of people think to look up at night," says Sonya Berger, an Acadia ranger. "They run around from one lit place to another lit place, flipping on artificial daytime whenever it gets dark. And then when you walk outside, usually there's going to be a streetlight or you're just going to walk to your car and flip your headlights on. It's kind of like living in Phoenix where you hop from one air-conditioned place to another, and you never feel the heat—you never realize that it's a really crazy place to live in during the summer. I think that for a lot of people that's what happens with the night sky, too."

In response, Berger and her fellow rangers offer a series of different programs geared to get park visitors to pay attention to the night, including Stars over Sand Beach, which on summer nights often has two hundred attendees or more. Knowing the Night, a night walk focused on sensory experiences, invites people to walk in the dark, testing their natural adaptations. It's not something many visitors have done before, the rangers find. Berger says, "We tell them it's okay, come with us."

It's a message Acadia National Park has been offering for decades, and a message the NPS as a whole is beginning to offer more and more. More than sixty national parks and monuments offer some form of night sky program, and the number continues to grow. In addition, the NPS as a whole has begun to take the conservation of darkness more seriously, adopting in 2006 a policy that seeks to "preserve, to the greatest extent possible, the natural lightscapes of parks, which are natural resources and values that exist in the absence of human-caused light." The policy also directs individual parks to "minimize light that emanates from park facilities, and also seek the cooperation of park visitors, neighbors, and local government agencies to prevent or minimize the intrusion of artificial light into the night scene of the ecosystems of parks." Since the policy change, several parks and monuments have taken the initiative to change their lighting—replacing old fixtures with more energy-efficient ones, shielding lights—and to begin encouraging an increased appreciation among visitors of the importance of night and darkness.

While the new policy has been embraced by some parks more than others, its message seems very much in line with the Park Service's original mission from 1916: "to conserve the scenery and the natural and historic objects and the wild life therein and to provide for the enjoyment of the same in such manner and by such means as will leave them unimpaired for the enjoyment of future generations."

There is no doubt that when light pollution erases the stars over a park, or interferes with the natural cycles of wildlife, or blights the view of mountains, waterfalls, or mesas, it impairs a park. Without action, the problem will only get worse.

America's national parks—and national parks around the world—represent an excellent opportunity for the preservation of darkness, and—I would suggest—the preservation of darkness represents a valuable opportunity for the parks. Any serious plan to protect and restore darkness, such as the Starlight Reserves or the IDA's designations, relies on a core dark area working with a buffer of surrounding communities. National parks already play this role in a number of issues, and would be well suited to serving as core areas for dark sky reserves. At the same time, if the parks themselves become hemmed in by a civilization insatiable for resources of every kind, eventually park boundaries will give way. Because darkness is affected by light from hundreds of miles away, it will always be one of the first natural resources to show signs of impairment by encroaching civilization. I remember Pierre Brunet arguing in Paris that the presence of an astronomer was the sign of a healthy ecosystem; that when the sky grows too bright for astronomy and the astronomers go away, you know you have a polluted sky, and whatever has polluted that sky will eventually pollute other resources, given time.

What makes Acadia such an important example for the national parks and darkness is the fact that it's a park with more than two million visitors every year, nestled next to several bustling communities. The park's dedication to darkness presents both opportunities to reach large numbers of Americans and challenges to protecting its dark skies. So far, it seems to be doing well on both fronts. In 2008, Bar Harbor citizens voted to enact a light ordinance that recognizes the importance of dark skies for the park and for the community, and in 2009 the park organized its first annual Night Sky Festival. Local businesses have joined in both efforts, recognizing that darker skies will mean more visitors to town. Berger says the park has received steady support from the community and believes local businesses are quickly realizing that the night sky as a natural resource is a tourist draw. "Because we're on the East Coast, and definitely in higher population areas, that relationship and the need for cooperation between the park and its partners is even more critical to achieve the park's preservation goals," she says. She believes those goals are important to the community as a whole. "Acadia National Park is almost a hundred years old," she explains, "and there's a long tradition of this being a place where people could go and enjoy the night sky."

"I had no idea about their sky," says astronomer Tyler Nordgren, author of *Stars Above, Earth Below: A Guide to Astronomy in the*

National Parks. "I just figured it was like any other place on the East Coast, where you'll see a smattering of stars in the sky, but that's it. Acadia is so beautifully isolated. You're up there on the Gulf of Maine. To go up to the top of Cadillac Mountain, or to drive the park road that winds along the coast, and to see the stars reflected in the ocean beside you—I was unprepared for it."

In writing his book, Nordgren spent every clear night he could outside at a national park looking at the sky, he says, trying to photograph it and write about astronomy in a way that would reach more people than just astronomers. In addition, he's been an artist since he was a child, a hobby he's increasingly embraced in his role as a teacher of science. "When you go out and you see a starry sky overhead, your first thought isn't for all the numbers and facts that go along with it. Your first thought is just how beautiful it is." He said he asked himself, "How can I show people in an evocative way, in an emotional way, just how wondrous it is to be out there? And you just can't help but do that through word and picture and art."

By art, Nordgren refers to the series of posters he's drawn advertising the night sky at Acadia and other parks, modeled after the Works Progress Administration (WPA) art of the 1930s. "I'd always loved those old WPA posters from the parks," he says. "And so I thought, Let's find a way to incorporate new posters of the night sky and parks and planets." His WPA-style posters are evocative, with small human figures standing in awe of a wondrous night sky. In a poster advertising "SEE THE MILKY WAY in America's National Parks," a man and a woman stand atop an outcropping of pale blue rocks, the Milky Way rising lava-lamp-like before them, the surrounding sky filled with white circles for stars. It is a scene of the sublime, of the small human figure facing the immensity of the universe. Seeing the poster, I think of author Bill Fox talking about the Black Rock Desert. "You know," Fox told me, "we think because of television, the Internet, or jet travel we see a lot of the planet. But the only chance we really have to retain our sense of the

An example of astronomer and artist Tyler Nordgren's WPA-style posters on behalf of the night sky. *(Tyler Nordgren)*

scale in the real universe is by looking at the night sky. And, to really see the night sky, it's like, 'Oh, my, it's a really, really, big universe out there.' "

"Thanks to art," Nordgren says, "I am able to do my science so much better than I ever thought I would." While obviously quite knowledgeable about his field, he is anything but "the learn'd astronomer" from Walt Whitman's nineteenth-century verse:

When I heard the learn'd astronomer;
When the proofs, the figures, were ranged in columns before me;
When I was shown the charts and the diagrams, to add, divide, and measure them;
When I, sitting, heard the astronomer, where he lectured with much applause in the lecture-room,
How soon, unaccountable, I became tired and sick;
Till rising and gliding out, I wander'd off by myself,
In the mystical moist night-air, and from time to time,
Look'd up in perfect silence at the stars.

Nordgren cringes when I mention Whitman's lines—the idea of an astronomer making someone "tired and sick" makes his stomach turn—but he acknowledges that reducing the night sky to dry numbers will often do the trick. He knows it isn't "the proofs, the figures" that will move his audience but the opportunity for each person to see for him- or herself, to wander off "in the mystical moist night-air" and look at the stars. It's the opportunity to do this that the national park system is so well-situated to offer to millions of visitors each year.

"When you're there at the South Rim of the Grand Canyon, and you're there on one of those smoggy days, and you don't have the pristine view of that canyon sweeping before you, anyone, even the least knowledgeable person, knows that they're missing something, that they're being robbed of something," he explains.

But, he says, when it comes to the night sky, most people don't know what they're missing.

"Everyone's grown up in cities. We have no idea that it can be any other way. People no longer realize that you should be seeing thousands of stars. You should have stars from the zenith to the horizon. People see that orange glow, the color of the sky back home, and think, Okay, well, maybe that's just the way the sky is."

At dozens of parks and monuments, dedicated rangers are doing everything they can to convince visitors that the sky they know back home is definitely not just the way the sky is. Through interpretive programs, astronomy festivals, full moon hikes, campfire presentations, and individual chats, NPS rangers talk about the practical dangers of artificial light and the steps visitors can take. But along with such practical advice, NPS rangers are sharing an idea as well.

In a 2007 documentary, filmmaker Ken Burns called the national parks "America's Best Idea," arguing that they protect not only scenic landscapes, funky oddities, and wildlife but also the intangible qualities that have made us who we are and who we might still be. As burgeoning human populations bring change to wild areas everywhere, the parks represent something that lasts. Though they may change in infrastructure or priorities, their core mission remains: to protect for Americans the opportunity to have an experience similar to what they—or other Americans before them—had in the past. In protecting their physical geography, the parks protect their intangible geographies as well. Preserving what NPS policy calls "lightscapes" goes right along with that idea. Increasingly, our national parks may represent our best chance at knowing, protecting, and restoring real darkness.

"We're at a delicate time now," says Tyler Nordgren, "where we still have some people that know what they're missing. But if we wait too much longer, everyone will have lost this. No one will realize it anymore. And it won't occur to anyone to want to preserve it.

"If we let another generation or two go away, we will have

largely lost that generation that said, 'I used to be able to see the Milky Way.' And then, once you've lost it, you've lost the major drive behind preserving it. Because there will be no one around anymore to want to make things go back to the way they were.

"So, now's the time."

In the Lower 48, nowhere east of the Mississippi is as dark as places west. Once you come across the great river, you can find geographies of darkness in western Nebraska and eastern Montana, in northeastern New Mexico and east-central Oregon. But for the most part a visitor to these remote areas would be on his or her own, without guidance or accommodation, and while that type of solitude has an undeniable appeal, it won't encourage large numbers of people to savor dark skies. More importantly, these areas are only dark by chance—the lights of civilization just haven't reached them yet. Especially because they are remote and rarely visited, their darkness is almost certainly doomed, as it will have few defenders when its inevitable enemies arrive. It's true with any other intangible resource: If few people care about the darkness that remains, eventually that darkness will disappear.

If we are ever going to protect the darkest places we have left, they will have to be dark places we actually know and visit, love and respect. As I reach the end of my journey, what makes most sense to me is to name not one single darkest place but a certain darkest geography, one blessed with dark locations to which people from all over the world already love to come. For me, the American Southwest, and especially our national parks and monuments there, represents the geography of night I've been looking for. There certainly are places in the world that are even darker, but no geography combines the darkest areas we have left with specific places we already love and have promised to protect. For my money, the national parks represent the best opportunity we have to experience, protect, and restore darkness.

What are some of these specific locations? For a while now, since the NPS Night Sky Team began making its readings, the darkest places have been these: Natural Bridges National Monument, Capital Reef National Park, and Bryce Canyon National Park, all in southern Utah. More recently, Big Bend National Park in southern Texas has joined the mix. After his months traveling the park system, Tyler Nordgren ranked the top five darkest parks he visited as Big Bend, Bryce Canyon, Natural Bridges, Grand Canyon, and Chaco Culture National Historical Park. But his rankings, he cautions, "reflect what I happened to see on those nights I happened to be there. They are by no means definitive." So many qualities go into making one location darker than the next, qualities that can change from night to night—among the most important are weather, season, and phase of the moon. A particular location may be the darkest one night, second darkest the next, third the night after that. And, in ten years, who knows what the list might look like?

Chad Moore explains. "Each year, they come up with what's the mean population center of the U.S., and it's always some town in Kansas, and it's kind of an artificial computer geographic operation. Similarly, we could say the darkest spot in the forty-eight states is right there in eastern Oregon or northern Nevada or some random place, but that doesn't carry the same sort of charisma. I think the point is not to find the light equivalent of 'one square inch of silence,'" he says. "I think what we need to do is to find these charismatic places that we already love where the night sky is woven into the fabric of that park, of that place, and then say, 'This is one of the darkest places—we need to defend this, we need to cherish this—and not worry about splitting hairs.'"

Here are two of those places.

In Bryce Canyon National Park, Ranger Kevin Poe stands with his back to the salmon-pink hoodoo columns of weathered rock that gave the park its fame, waving his hand toward the horizon. "Ladies and gentlemen," he says to a group of park visitors, "on behalf of the National Park Service, I present to you...the full moon." Moments later, the moon climbs into the sky. The knowing people smile, but to the casual observer who might not know that astronomers can gauge the moonrise (and moonset, as well as sunrise and sunset) to the exact minute, and centuries in advance, Poe's timing must almost seem magic. For Poe, a big man with a long ponytail and huge love for the night, the event simply marks the start of another full moon hike at Bryce, another chance to help visitors fall in love with the dark.

As he leads the group of twenty-five visitors (many of whom, Poe says, have never heard the words "light" and "pollution" in the same sentence before), he mixes constant good humor into a wide-ranging presentation that includes the sounds of night, the importance of bats, and the threats from "the reckless addition of more and more light to the sky." At the bottom of the trail, where the tour turns around, he and I wait, then bring up the rear. "It's all part of helping them understand that Bryce Canyon does not disappear at dusk," he explains. Not only does it not disappear, but in many ways it's after dark that Bryce best shines. On these ever-popular full moon hikes—the tickets become available at 6:30 a.m. and are gone within an hour—rangers lead visitors through a park that is one of the darkest in the nation. "We used to say we're a one in the wintertime and a two in the summer," Poe explains, referring to the Bortle scale. "But the winter is frustrating now because we have a lot of jet contrails"; the winter sky, while still exceedingly dark here, is no longer as clear.

As we climb back up the wide graded trail, past a soft breeze in the pine-tops, trunks and needles radiating moonlight, I tell Poe about what I've been looking for—dark places, but not so remote that no one can get there. "We're not pristine anymore," he says, "and Chad will be the first person to say that probably no longer exists. But we're pretty close, and I think it's fair to say we're as close as you can drive to."

Poe stops to point out a flower blooming in the moonlight.

"The bronze evening primrose," he says. "Pollinated at night by sphinx moths." On my hands and knees I lean in to inhale the wonderful scent, a scent that lingers as we continue up the trail. "My secret best dark location is—" (It is a secret I cannot repeat.) "My boys and I do these marathon canoe trips to get there. It's one of those places where nobody's going to be able to drive to but it's very Class One-ish, and, boy, it's an extraordinary place." When I tell him that I'm headed next to Natural Bridges, he smiles. "Well, Natural Bridges is in that same spot when you look at one of these NASA images, and that's one you can drive to."

A few nights later I find myself sitting alone on a pile of massive rocks, waiting to meet the darkness of Natural Bridges National Monument. In 2006, the IDA named Natural Bridges its first International Dark Sky Park. The NPS Night Sky Team had ranked the level of darkness as 2 on the Bortle scale, and as Chris Luginbuhl explained, "Basically, that means it's the darkest or starriest sky they've seen while doing these reviews."

Kevin Poe is right: You can drive to "the darkest or starriest sky they've seen." You can park in paved parking, use the clean basic outhouse, walk the paved path to the overlook. I'm not saying you have to—there are backcountry options as well—but don't think just because many consider this the darkest place in the National Park system that you can't get here easily. True, the drive takes a while, and if you come the way I did, you climb the side of a canyon that will have you wondering about a letter to the maker of your rental asking if their new cars ever stall and start to slide backward. But when you get here you may find it almost deserted if it's a weekday early in the season, with hardly anyone in the campground and no cars on the loop road. You can park and walk out to this pile of rocks, climb up easily, and wait for night.

Three large natural stone bridges cross over a curving canyon full of dark piñon green, red rock cliffs for a backdrop. To be where anyone else could be and yet to find yourself alone feels like discovering a secret. So you're out here on this big rock, waiting for darkness to rise around you, shoes off, your feet bare to the breeze. And here's what happens: The longer you stay, the darker it gets, and the darker it gets, the more sounds arise—crows and frogs in the canyon, crickets all around, and sounds that make you think of lions. Despite Kevin Poe's having said that you're more likely to be killed by a vending machine than a mountain lion, you can't help but feel that tingle of fear, that fear of the unknown, that mystery. You like the feel of bare feet on warm desert rocks, the unexpected scent of night-blooming rose. You lie on your back with your hands across your eyes like blinders, making the world that much darker, then open them to reveal the sky. You do this again and again, and each time the sky is a little brighter, each time more peppered with stars. You stand and open your arms, savoring this window of darkness between the end of twilight and the waning moon's rise. You feel the breeze on your skin and in your hair, hear the sounds from the canyon of crickets and crows and the steady throb of some creature unknown; you feel utterly surrounded by natural night, by fellow creatures for whom this is home, none of whom care if you're here as long as you don't bother them, all of whom lend their voice to the song this night sings, saying wel-come wel-come wel-come, belong.

And here are two more, both personal, places I hope we all share.

The first is memory, mine of the night in Morocco when I thought I'd stepped into a snowstorm. I thought, when I began this book, of trying to go back, of trying to find that location and maybe even that sky. But instead I decided to protect that memory, and to look for similar nights elsewhere. This is the place of our firsthand experience with night—beautiful, inspiring night—that I've heard about again and again from those I've met, and that forms the basis for any future concern about darkness. The opportunity to experience a real, dark night, especially when we are young, imprints on our minds a vision we never lose, one we might be inspired to regain.

The second is a night called home, for me a lake in northern Minnesota. These days I'm only there during summers and sometimes at New Year's. But this is the night that means the most to me, the one that moves me to act. If we are going to protect darkness, we almost certainly will do so because of the darkness we cherish or wish to see again in the place we call home. Just as Edward Abbey wrote, "This is the most beautiful place on earth. There are many such places," so I see the night at the lake. Even if it's no longer pristine, it's the most beautiful night I know, the night I want to protect most.

And finally, one more, the dark place with which I'll close, the darkest place I have visited, is one where I am not alone, and this makes all the difference. For the darkness I believe so valuable won't be protected and restored only by solitary citizens going out into the night, but rather by places like this on nights like this—when dozens of children join dozens of adults under a breathtaking starry sky.

Dan Duriscoe's autumn 2005 report on Great Basin National Park reads like an astronomer's dream: "Airglow has distinct blue green color, *gegenschein* easily seen but not the entire zodiacal band. Detail in the Milky Way in Cassiopeia substantial, M33 easy naked eye object, seen with direct vision. Light domes of Las Vegas and Salt Lake City are apparent but not brighter than Mars. Would be Bortle class 1 or 2 if not on high mountaintop." In other words, except for a few spots on the horizon, the sky over Great Basin National Park is as dark today as it was before European settlement, so dark that a natural glow ("air glow") hovers in the air, and the "*opposite shine*" lights the sky opposite to the sun. Even a nearby galaxy (M33) can be seen with the naked eye. The only drawbacks to this spectacular site are—as in Death Valley—the faint light domes of distant cities, its location at the summit of one of the park's mountains, which, Duriscoe admits, makes it "very exposed, cold, windy"; and that "lack of oxygen," he notes, "may be a handicap for visual observing."

Thankfully, the night after my drive here, playing with my headlights under the Milky Way, I stand not on a mountain peak but in the picnic area near the park's visitors center, surrounded by two dozen amateur astronomers, their forty-plus telescopes, and most of the nearly three hundred other visitors who have come to the park for its annual Astronomy Festival. And what a diverse group we are—grandparents seated in folding chairs, mothers and fathers and excited kids, young backpackers in dirty boots and shorts. The park had its first festival in 2010 and in one night saw its visitation numbers spike. This year, it has expanded the festival to three nights, and still the campgrounds are full and the parking lots packed. Earlier, at a ranger-led program at sunset, many of these visitors gathered to read poems and sing songs inspired by the night, beginning with three small children singing "Twinkle, Twinkle Little Star" and ending with a ranger-led sing-along of "Home on the Range," though most of us could only mumble the little-known second verse:

How often at night when the heavens are bright with the light from the glittering stars,
Have I stood there amazed and asked as I gazed if their glory exceeds that of ours.

Now, not long after the program ended, standing "amazed" is what most of us are doing. To the south, west, and north the horizon ends with a close mountain's rise. To the east, the Snake Valley runs for miles into Utah before reaching the mountains, where Jupiter hangs like an untethered balloon, lit from within, glowing as though lifted by coals. When a particularly bold shooting star passes overhead leaving a trail of smoke, a collective "Ooooohhhh" rises instinctively from those in the crowd who saw it, and good-natured curses from those who were looking the other way. It's fair to say that few of us know one another, and yet fair as well to acknowledge a feeling of community here as we share the night's experience. All around, those stars nearest the horizon are shimmering as though in a breeze and seem brighter, and even somehow larger, than they ever appear over the city. Their colors, like the red of Antares beating at the heart of the constellation Scorpius, are more evident than ever. "I'd forgotten how long it's been since I've seen stars like this," says a woman nearby. "I've never seen stars like this," says her younger friend.

I have seen stars like this, but not often. And I wonder tonight at how rare this experience has become. While writing this book, I have often wondered how hard anyone should have to work to simply see a truly starry sky, to simply know a truly dark night. In months of travel, getting outside every chance I could, I have had only a few nights like this one. The weather is cooperating, the moon is off doing other things, and there isn't some strange natural disaster messing up what everyone told me the sky was normally like. The *calima* in the Canaries was the worst example of this, but smoke from the largest forest fire in the recorded history of the Southwest clouded many of my nights, and more than one astronomy festival "star party" was crashed by weather everyone told me was odd. I have often felt as though in paying close attention to different levels of darkness I have been seeing a changing climate firsthand. As if "lights and ever more lights" weren't enough, we also now have this.

"What an incredible experience for the kids," says my neighbor. Her friend replies, "Everybody wants their kids singing 'Twinkle, Twinkle' and knowing what it means." Remarkably, estimates are that eight out of every ten children born in America today will never know "what it means." That is, 80 percent will never know a night dark enough that they can see the Milky Way. Standing under this clear desert sky, a statistic like this seems wholly implausible, as though reporting that eight of ten children will never speak or run or go outside. And yet, for every child out here tonight there are many thousands back in Las Vegas living swamped in light, without even the opportunity to "wonder what you are." The words to the well-known children's song come from a book of British nursery rhymes published in 1806, a time long before electric lights, a time when,

Then a traveler in the dark,
Thanks you for your tiny spark,
He could not see which way to go,
If you did not twinkle so.

In a world where a star's light could guide a traveler's way, any child would be drawn to wonder. And not only children, but the rest of us, too. Henry Beston wrote, "When the great earth, abandoning day, rolls up the deeps of the heavens and the universe, a new door opens for the human spirit, and there are few so clownish that some awareness of the mystery of being does not touch them as they gaze." Tonight's sky causes me to think that, even with what can sometimes seem such overwhelming obstacles in our future, Beston's words still hold true: Given a chance to be touched by the beautiful mystery of night, there are few of us who won't feel our spirit strengthened, our will resolved. The lights from Vegas may not stay in Vegas, but neither will an experience like this, an experience available still in this geography of night, and one that will return with us to wherever we call home.

Beneath this Great Basin sky, that kind of inspiration comes easily, and thoughts instinctively turn to one's place in the world, and the world's place in the universe. This feeling, of tilting your head back until you feel enveloped by stars, of wonder and wondering, feels as primal as my experience last night driving to the park, that sensation of being thrown from the edge of the world.

"There's a name for that," says Bill Fox, the festival's keynote speaker and my fellow stargazer tonight. "When the horizon disappears and you feel like you're falling into the stars, it's called 'celestial vaulting.'" Fox tells me of an artist named James Terrell who has already spent $23 million building a piece of art out of an extinct volcano's crater outside Flagstaff with the explicit goal of creating for viewers the experience of "celestial vaulting." And this experience of coming face-to-face with so many stars—when the sky opens as though a vault and we feel we are falling—is an experience that matters, Fox explains, because "if we never see the Milky Way or feel ourselves staring into the surrounding universe, how can we really know where we are? How will we know our place in the universe?"

The author of more than a dozen books, Fox has long been fascinated by the way the human mind struggles to make sense of where we are, especially when faced with large spaces—the night sky included. Earlier, he'd told me of how American bomber pilots flying night missions in World War II found that for weeks after the end of their tour they could not focus on distant objects—the result of long hours of staring out into space. Their eyesight checked out fine, Fox explained, but their mind had lost the ability to make sense of what their eyes had been straining to focus on.

Fox grew up in Reno in the midsixties when he could see the Milky Way from his front door, and the city cops who were initially suspicious of his front-yard telescope were soon coming regularly to check out his view of the heavens. In work that has focused on geographies like Nevada's Great Basin, the Australian outback, and the white "deserts" of the Arctic and Antarctic, Fox has seen that view steadily fading, even, he has written, in the Arctic: "My Inuit friends have been saying for several years that the night is no longer as dark as it used to be. But no one believed them until the local meteorologist discovered that a layer of the Arctic atmosphere, recently warmed by global climate changes, was reflecting sunlight from far below the horizon. So even the polar night, the longest and most pure form of that black isotropy we find on earth, is threatened by the ubiquitous footprint of our species."

Fox winces as the headlights from a car leaving the visitors center flash across the picnic grounds, momentarily blinding us. After an hour of wandering from telescope to telescope, following green laser pointers wielded by silhouetted astronomers reeling off the Arabic names of the stars, and enough stargazing that our necks are growing tired, our eyes have grown used to the dark. "Makes you realize how bright those lights are," Fox says, "lights that in the city we're so used to that we don't even notice."

I nod, thinking of the Wendell Berry poem I have carried with me while writing this book:

> *To go in the dark with a light is to know the light.*
> *To know the dark, go dark. Go without sight,*
> *and find that the dark, too, blooms and sings,*
> *and is traveled by dark feet and dark wings.*

How upside down this world where what was once a most common human experience has become most rare. Where a child might grow into adulthood without ever having seen the Milky Way and never feel as though lifted from Earth into surrounding stars. Where most of us go into the dark armed not only with "a light" but with so much light that we never know that the dark, too, blooms and sings.

How right it feels to be in this place, standing with dozens of others, gazing at the Milky Way. How right it feels to know a true night sky, how right to know the dark. And as my companions and I head back toward the parking lot, back toward the light, I let the others walk ahead, and turn—one more time before I go inside, before the lights take my night vision—to see in that darkness our home in the universe, the rising ribbon of billions of stars sashed overhead, horizon to horizon, just as it always has been.

Acknowledgments

This is something I have been looking forward to for a long time: the chance to say thank you in a public way to so many who have helped me write this book.

Let me begin with lifelong friends, friends with whom I have discussed this work many times or who have influenced the writing in some way. First, to Thomas Becknell, whose thoughtful words have inspired me for the past ten years. To Emily Spiegelman, a friend since our days in New Mexico. To Ingrid Erickson, since Carleton. To Christina Robertson, in Reno. And to Tiffany (Threatt) Bourelle. To Randall Heath, for thirty years of conversations and laughs. To Marty Huenneke, for his essential company in Spain and a brogue that always brings a smile. To Eric Stottlemyer, from Reno to Winston-Salem and beyond. To Joshua Powell—we'll always have "the gold," my friend. And to David Swirnoff, whose wisdom and humor have kept me afloat from the days we were teenagers running around Lake of the Isles at midnight to his trusted reads of this manuscript.

To longtime family friends who have supported me along the way: Marjorie Bjornstad in Milwaukee, Susan Flint and Michael Leirdahl in Minneapolis, Anne and Jack Ransom in Minneapolis, Mary and Jack von Gillern at Thunder Lake, Kathleen and Gene

Scheffler in Golden Valley. Jeanne Harrie and Jerry Kleinsasser in Bakersfield.

To Carly (Johnson) Lettero, Tom Schmiedlin, Michael Macicak and Carmen Retzlaff, Patrick Thomas, Michael Leville, Andrew Comfort, Alison Van Vort, Rachel and Joel Crabb, Nancy and Ron Crabb, Jim Barilla. To Scott Dunn, from editing my work in Albuquerque to housing me while I replaced La Rosa. To Douglas Haynes for hospitality in Madison and emergency cleaning services the day the shaving cream exploded in the front seat.

Friends from Carleton College days: Bardwell Smith, Wendy Crabb, Laura and John Gibson, Kristin Tollefson, Hanna Cooper, Laura (Kindig) Timali, Stephanie Satz and Jeremy Alden, Scott Dale.

From Albuquerque: Bobbo McCormick, Gordon Schutte, Dan O'Brien, Bonnie Nuttall, Adam Ford, Cara O'Flannigan, Blake Minnerly, and so many Albuquerque Ultimate friends. To Derek Sundquist (Go Gophers!), and in memory of Bailey. To Rachel (Armenta) Menke. And with special thanks to Greg Martin, a wonderful teacher of creative nonfiction.

From Reno: Jen Hughes Westerman and Jim Westerman, Mike Branch, Cheryll Glotfelty, Chris Coake, Andy Burelle, Heather Krebs, Lisa Fleck, Kyle Ferrari, Amy Poetschat, Rich and Jackie Starkweather, Jim Frost, Matt and Katie Anderson, Sudeep Chandra, Megan Kuster, Leslie Wolcott, Dawn Hanseder, Justin Gifford, Dan Montero, and all my friends at Reno Ultimate. To 535 Toiyabe Street, the Bibo Coffee Company, and the trails behind Patagonia.

Special thanks to Scott Slovic for his advice, ideas, and unfailing optimism.

I am thankful to have spent three years teaching at Northland College in Ashland, Wisconsin, where I worked with an extraordinary group of faculty that included Erica Hannickel, Paul Schue, Jason Terry, Michele Small, Tim Ziegenhagen, Tim Doyle, Elizabeth Andre, Alan Brew, and Grant Herman. What would life be like without my knowing Cynthia Belmont? I cannot imagine.

David Saetre is a man whose friendship I treasure. Mary Rehwald helped make 715 Ellis Avenue feel like home. I am honored to call a true international man of mystery, Dr. Rajat Panwar, my friend.

For two years at Wake Forest University it was my pleasure to work with many fine colleagues, including Jessica Richard, Dean Franco, Erica Still, Rian Bowie, Eric Wilson, Scott Klein, Ryan Shirey, Collin Craig, Anne Boyle, Laura Aull, John McNally, Grace Wetzel, Patrick Moran, Rachel Deagman, Mary DeShazer, Cynthia Gendrich, and Phoebe Zerwick. Thanks to Kendall Tarte for making an important call to Paris on my behalf, and to Bill O'Connor for showing me his moth collection. Special thanks to Miles Silman for including me in his work with the Center for Energy, Environment, and Sustainability. To Erin Branch and Lukas Brun in Chapel Hill, thank you for the wonderful dinners. To Omaar Hena and Gretchen Stevens, I will not soon forget the text that simply said (and said it all), "Do you prefer salmon or filet mignon?" and signaled another night of great food and wine on the back deck. To Abi Flynn, for indulging my endless enjoyment of her English accent and sharp wit.

I am grateful for the grants I received while at Wake Forest University, which supported my research for this book. These include grants from the Archie Fund for Arts and Humanities, the Dingledine Fund, and the Center for Energy, Environment, and Sustainability (CEES).

To my new colleagues at James Madison University, thank you for bringing me aboard. Thanks especially to Laurie Kutchins.

In addition to offering my deepest thanks to those I quote directly, I wish to thank several others who were instrumental in my writing and research. These include Lynn Davis of the National Parks Conservation Organization, James Fischer of Zoolighting, Roberta Moore, Kelly Carroll, and the rangers at Great Basin National Park, Peter Lipscomb in Santa Fe, Kathleen Dean Moore of Oregon State University, Don Miller of Severson Dells

Nature Center, Mary Adams from the Headlands Dark Sky Park in Michigan, Neil deGrasse Tyson of the Hayden Planetarium, Gary Harrison of the University of New Mexico, Siegrid Siderius in Amsterdam, Nicolas Bessolaz in Paris, Wim van Driel in Paris, Belgium's Friedel Pas, Franz Hölker in Berlin, Rowena Davis and Scott Kardel of the International Dark-Sky Association. Alex Pollard in London, Yves and Sandrine Lavenant in Paris, Alison Harris in Paris, Bob Crelin for the drawings, Peter Baldwin for his book, Paul Klass for the legal background. In addition, here's to many more wonderful meals in Paris with Emma and Philippe Aronson.

Special thanks to Christian Luginbuhl, Richard Stevens, and Steven Lockley for their expert knowledge and interest in this book.

I remember the day I heard my agent's name for the first time. I was walking on the Northland College campus with Steven Rinella, a visiting writer from New York, telling him about my book idea. "Oh," he said, "you gotta talk to Farley." Farley Chase has been everything a writer could hope for in an agent, and I look forward to working with him for years to come.

To have this book published by Little, Brown is a dream come true. I have had the great pleasure of working with John Parsley as my editor. His ever-present cheerfulness and wise editorial eye have made this book significantly stronger than it would have been.

Thank you to everyone at Little, Brown for working hard to bring this book out into the world. Thanks especially to Pamela Marshall and to Carolyn O'Keefe, and to Janet Byrne, my copyeditor. Thanks to Tyler Nordgren for the cover art.

Thank you to Louise Haines at HarperCollins UK.

Finally, to my family: aunts Joan, Myrna, Mary, and Ruth; Uncle Jim and Aunt Carol; my cousins; my sister, Rachel, and brother-in-law, Bob. To the memory of my grandparents, Cecil and Evelyn Bogard, and Milton and Gladys Holcomb. To Luna, the best dog friend ever. And to my parents, Judith and John Bogard, who have been there for me always.

Notes

Introduction

In Isaac Asimov's story "Nightfall" (1941), six suns surround a planet so that it never experiences darkness. When a fluke eclipse blocks all six suns, civilization is thrown into panic.

The advertising slogan "What Happens Here, Stays Here" has been used by the Las Vegas Convention and Visitors Authority since 2005.

The images I describe from the 1950s, 1970s, 1990s, and 2025 come from The World Atlas of the Artificial Night Sky Brightness, created in 2001 by Pierantonio Cinzano and Fabio Falchi at the University of Padua in Italy. Cinzano and Falchi took mid-1990s satellite data provided by Chris Elvidge at the National Oceanic and Atmospheric Administration (NOAA) and created their colorful maps by estimating backward and forward in time. They are now working on a new atlas based on more recent data. The original atlas may be viewed at: http://www.lightpollution.it/worldatlas/pages/fig1.htm.

The information on Mizar and its fainter binary, Alcor, comes from Emily Winterburn's entertaining *The Stargazer's Guide: How to Read Our Night Sky* (New York: HarperCollins, 2008). She defines Mizar as a visual binary, meaning "pairs of stars that revolve around each other…due to the gravitational pull exacted on one by the other." Arab astronomers knew Mizar and Alcor together as "horse and rider."

Wendell Berry's poem "To Know the Dark" comes from *The Selected Poems of Wendell Berry* (San Francisco: Counterpoint, 1998).

The attraction of squid to bright lights is much like that of moths to a flame. For more on squid boats and their day-mimicking lights, go to Russ Parsons, "Lights, Nets, Action," *Los Angeles Times*, January 31, 2007, http://www.latimes.com/la-fo-squid312007jan31,0,5288418.story?page=1. A squid fishing boat light is "100–1,000 times brighter than the natural condition," explain researchers at Japan's Hokkaido University (www.pices.int/publications/presentations/PICES_12/pices_12_S3/Fujino_956.pdf), in an article that includes a dramatic view of squid fishing fleets as seen from satellites. Also see "Bright Lights, Big Ocean" (http://www.darksky.org/assets/documents/is193.pdf), which details the effects of cruise ships and offshore oil rigs and argues, "it is not true that the darkness of the oceans is like it was even 20 years ago." And this was 2003.

For an excellent history of early lighting "technologies," see Jane Brox's *Brilliant: The Evolution of Artificial Light* (New York: Houghton Mifflin, 2010). In just one example, "Shetland Islanders caught, killed, and stored storm petrels by the thousands." When islanders needed a torch, they would affix a petrel carcass, "a sea bird, full of buoyant, insulating oil...to a base of clay, thread a wick down its throat, and set it alight."

In the February 2001 issue of *Sky & Telescope*, John E. Bortle explained the origins of his scale this way: "Unfortunately, most of today's stargazers have never observed under a truly dark sky, so they lack a frame of reference for gauging local conditions. Many describe observations made at 'very dark' sites, but from the descriptions it's clear that the sky must have been only moderately dark. Most amateurs today cannot get to a truly dark location within reasonable driving distance....To help observers judge the true darkness of a site, I have created a nine-level scale."

"Thirty years ago," Bortle lamented, "one could find truly dark skies within an hour's drive of major population centers." But "this is no longer possible."

The winter I really began to learn the constellations and feel a hunger to know more about the night, the night sky, and darkness, I spent several evenings on the floor of the (now defunct) Border's Bookshop in Albuquerque, paging through every book on the shelves marked "astronomy." No book caught my interest more than Chet Raymo's *The Soul of the Night: An Astronomical Pilgrimage* (Lanham, MD: Cowley Publications, 1992; reprint, 2005). Mixing analogies ("If the sun were a golf ball in Boston, the Earth would be a pinpoint twelve feet away, and the nearest star...would be another golf ball...in Cincinnati") with quotes from Henry David Thoreau, John Burroughs ("The gifts of night are less tangible"), Sylvia Plath, Rainer Maria Rilke, Theodore Roethke ("In a dark time the eye begins to see"), and many others, Raymo captured the sense of fascination and wonder that so often accompanies becoming aware of the world after dark.

Because we can see only a fraction of the universe, we can only guess at the number of galaxies it holds. Author Fraser Cain writes, "The most current estimates guess that there are 100 to 200 billion galaxies in the Universe, each of which has hundreds of billions of stars. A recent German supercomputer simulation put that number even higher: 500 billion. In other words, there could be a galaxy out there for every star in the Milky Way" (http://www.universetoday.com/30305/how-many-galaxies-in-the-universe).

In Greek and Roman mythology, the Milky Way formed from Hera's breast milk (Greek) or from the breast of Ops, Opis (Roman). But other ancient cultures had different explanations. One Cherokee folktale tells of cornmeal spilled by a thieving dog and describes the Milky Way as "The Way the Dog Ran Away." People living in southern Africa's Kalahari Desert saw it as embers from a fire. Australian Aboriginal people saw it variously as a river in the sky-world, termites blown into the sky, and thousands of foxes carrying away a dancer. Whatever the explanation, for every culture across the world the Milky Way was a regular presence in everyday life.

9: From a Starry Night to a Streetlight

"I am sure Las Vegas has the brightest pixel of any urban center," Chris Elvidge told me from his office at NOAA in Boulder, Colorado. In 1996, Elvidge used data gathered from a military satellite 528 miles overhead to map city lights and determined that Las Vegas was the brightest city on Earth, with New York and Madrid runners-up. While nearly two decades of rapid economic growth in China has made many of the cities in that country quite bright, Elvidge told me that "based on the Luxor beacon" Las Vegas could still claim the brightest spot on Earth.

Of his "Class 9: Inner-city sky," John Bortle wrote: "the entire sky is brightly lit, even at the zenith. Many stars making up familiar constellation figures are invisible, and dim constellations such as Cancer and Pisces are not seen at all. Aside from perhaps the Pleiades, no Messier objects are visible to the unaided eye. The only celestial objects that really provide pleasing telescopic views are the Moon, the planets, and a few of the brightest star clusters (if you can find them). The naked-eye limiting magnitude is 4.0 or less." This describes the sky over Las Vegas, New York, and dozens of cities around the world. But this tells us what we still can see. What has been lost? Consider that in a Class 9 sky the winter constellation Orion shows only its brightest stars, such as Betelgeuse and Rigel and the stars of its belt. Then, consider that these stars are brighter than 98 percent of the stars we should be able to see. That is, 98 percent of the sky has been wiped from view.

If the term "star party" conjures a vision of geeky astronomers ooo-ing and aah-ing while crowded around a telescope, well...that's often close to the truth. A star party can mean anything from a couple of amateur astronomers setting up their scopes on top of a campus building to multiday festivals that draw devoted stargazers from around the region, the country, or even the world to an especially dark location. In most cases, star parties provide people with excellent opportunities for public viewing alongside amateur astronomers who are more than willing to share everything they know about the sky.

In North America and western Europe, the Milky Way bends overhead twice a year, once in the winter and again in the summer. In the northern hemisphere we look toward the center of the galaxy during the summer, and so the view overhead is more dramatic than in winter, when we look away from the center of the galaxy.

While in Las Vegas their brilliant glow may blend in with the surrounding wash of light, in most locations digital billboards stand out with blinding force. Unheard of only a decade ago, these fabulously bright and ever-changing billboards have been spreading across the United States with remarkable speed: While as of 2010 only about two thousand of the country's four hundred fifty thousand billboards were digitized, several hundred are being added every year, and experts predict that there will eventually be more than fifty thousand. While proponents tout their ability to host several advertisers at once, critics call them "TV on a stick," claim they create safety hazards by distracting drivers, and warn that, once in place, they are very difficult to remove.

First observed in Sydney, Australia, in 2006, Earth Hour has now spread across the globe. While primarily intended to draw world attention to energy use and climate change, the movement's symbolic turning off of the lights on famous landmarks serves as a powerful reminder of our ability to address light pollution. See www.earthhour.org for details, including inspiring videos of the lights on landmarks around the world switched off, including the Eiffel Tower, the Coliseum in Rome, and the Opera House in Sydney.

Find Ellen Meloy's wonderful essay in *Raven's Exile: A Season on the Green River* (New York: Henry Holt, 1994) as well as in several anthologies. At the time of her sudden death in 2004, Meloy was at the height of her writing powers, having produced such books as *The Anthropology of Turquoise* (2002) and *The Last Cheater's Waltz* (1999).

The drawing of arc lighting on the Place de la Concorde in Paris that I'm thinking of can be found in Wolfgang Schivelbusch's *Disenchanted Night: The Industrialization of Light in the Nineteenth Century* (Berkeley: University of California Press, 1995). This fantastic study of the development of artificial light covers—with subtle humor and steady insight—the way artificial lighting changed from oil to gas to electric, and the effect of this change on the city street, the interior of the house, and the stage. For anyone interested in lighting at night, this book is invaluable.

Because arc lights were far too bright to be placed in existing gas lamp fixtures, they had to be placed high above the city streets. The latter part of the nineteenth century saw American cities as diverse as Denver, Los Angeles, Minneapolis, Mobile, San Francisco, and Buffalo erect tall towers topped with arc lights. For the most part, these towers failed to live up to their promise and quickly fell out of favor. But visitors to Austin, Texas, can still see several of the old (from 1895) "moonlight towers" in action.

The idea of turning night into day has never really gone away. Most recently, the American politician Newt Gingrich was ridiculed for his proposal (originally made in 1984) to position giant mirrors in space that would reflect sunlight down onto Earth and as *Time*'s Jeffrey Kluger explains, thus "eliminate the need for nighttime lighting on highways and serve as a deterrent to crime by brightening shadowy neighborhoods." See Jeffrey Kluger, "The Silly Science of Newt Gingrich," *Time*, December 15, 2011, www.time.com/time/health/article/0,8599,2102471,00.html#ixzz1mPsa9kJb.

Jill Jonnes's book *Empires of Light: Edison, Tesla, Westinghouse, and the Race to Electrify the World* (New York: Random House, 2003) is a well-told history of a changing world, "dramatically transforming man's age-old sense of day and night."

For a fascinating look at America before the advent of electric light, see Peter C. Baldwin's *In the Watches of the Night: Life in the Nocturnal City, 1820–1930* (Chicago: University of Chicago Press, 2012). From its first line ("To step into an unlit city street in early America was to enter a world shockingly different from our own") through sections such as "The Night Children" and "Sex and Danger in the Man-filled City," Baldwin's history describes for the contemporary reader this "shockingly different" world of not all that long ago. In an earlier article, "How the Night Air Became Good Air," Baldwin tells an amusing story about John Adams and Benjamin Franklin bunking together while on the road, and of Franklin wanting the window open but Adams wanting it closed for fear of the ill health night air would bring ("I, who was an invalid and afraid of the Air in the night"). Franklin's lengthy explanation to his friend that there was nothing to worry about eventually put Adams to sleep (*Environmental History* 8, no. 3 [July 2003]).

The drawing from John Jackle's *City Lights: Illuminating the American Night* (Baltimore: Johns Hopkins University Press, 2001) is on page 46. One of Jackle's most compelling arguments is that, above all, "public urban spaces at night were reconfigured to accommodate the automobile." The briefest look at any American city at night would seem to confirm this.

There is no shortage of information available about light pollution. See especially Bob Mizon's excellent *Light Pollution: Responses and Remedies* (London: Springer, 2002) and the International Dark-Sky Association's *Fighting Light Pollution: Smart Lighting Solutions for Individuals and Communities* (Mechanicsburg, PA: Stackpole Books, 2012), as well as the IDA's website, www.darksky.org. For a quick gauge of sky quality, locate the Little Dipper. If you can see all four stars in its cup, you have basically good, dark skies. If you can only see the two stars at the front of the cup, your skies are fair to poor. In most American and European cities you can't see the Little Dipper at all.

Examples of good (that is, fully shielded) and not-so-good (glaring) lighting fixtures. *(Bob Crelin/IDA)*

The quote from Thoreau about wanting to know "an entire heaven" comes out of his journal from March 23, 1856. See *The Journal of Henry David Thoreau, 1837–1861* (New York: New York Review Books, 2009).

Bob Berman's column on stupid questions, "'F' in Science," can be found in *Astronomy* from September 2003. My favorite: "Does Mars Have a Sun

Like Our Sun?" Berman's wonderful *Secrets of the Night Sky: The Most Amazing Things in the Universe You Can See with the Naked Eye* (New York: HarperCollins, 1995) has twenty-nine chapters divided among four seasons. His most recent book is *The Sun's Heartbeat* (New York: Little, Brown, 2011), and his website is www.skymanbob.com.

The quote from Michael Hoskin about the fascination with sheep guts can be found in *The History of Astronomy: A Very Short Introduction* (Oxford: Oxford University Press, 2003). For anyone interested in the subject (of astronomy, that is), Hoskin's book is an excellent choice.

The information on Julius Schiller's ill-fated attempt to Christianize the constellations comes from Winterburn's *The Stargazer's Guide*, one of the many books on viewing the night sky currently available. Perhaps one hopeful sign amid the ever-growing level of light in our nights is the fact that new books on viewing the stars are being published regularly—it seems that, even if we can no longer see the stars, we remain interested in learning about them. Among the many fine books on my shelf: *The Starry Room: Naked Eye Astronomy in the Intimate Universe*, by Fred Schaaf (New York: Wiley, 1988); *The Starlore Handbook: An Essential Guide to the Night Sky*, by Geoffrey Cornelius (San Francisco: Chronicle, 1997); and, of course, Chet Raymo's *An Intimate Look at the Night Sky* (New York: Walker, 2001). In addition, several smartphone apps offer instant information about the night sky. These include Pocket Universe and Star Walk for iPhone users and Google Sky Map for Android users. Of course, future apps will certainly be even more advanced—and the present ones are already impressive. But it's worth noting that these apps work whether your night is truly dark or not, that even if you can't see the stars, you can see the "stars." One wonders if using such apps will inspire a desire to know a real night sky, or if we will be content knowing what that sky ought to look like, if only.

For a fascinating exploration of Van Gogh's night paintings, see "The Skies of Vincent van Gogh," by Charles A. Whitney (*Art History* 9, no. 5, September 1986). By figuring out how the night sky would have appeared over Arles, France, in 1888 and consulting weather records, Whitney was able to determine the sky Van Gogh would have seen during the time he painted his famous works. Whitney then concludes that, of Van Gogh's three most famous paintings of the night sky (*Starry Night*, *Café Terrace at Night*, and *Starry Night on the Rhone*), the Big Dipper in the latter is the only clearly identifiable constellation, and even this has been transposed by the Dutch artist from the northern sky to the southeastern sky. Of the famous *Starry Night*, Whitney argues that the painting was done in Van Gogh's cell from memory, and that "he assembled his own sky from impressions gathered over an interval of a month or so." For anyone who appreciates Van Gogh's work, and especially his night paintings, a trip to Arles is rewarding. The famous café has kept its brilliant yellow paint, the exact point where Van Gogh painted *Starry Night on the Rhone* under gas lamps is marked (ironically, glaring white lights now clot the scene), and while the asylum in St.-Rémy where he painted *Starry Night* can only be visited during the day, his room has been reconstructed and preserved, and anyone familiar with the painting will instantly recognize the sloping hills in the distance.

Giacomo Balla was so taken with electric light that, in addition to painting his radiant *Street Light*, he named two of his three daughters Luce and Elettricità: light and electricity. He named his third daughter Elica, which translates to English as "propeller."

Joachim Pissarro, the great-grandson of the famous French painter, was curator at MoMA for the 2009 exhibition Van Gogh and the Colors of the Night. In the book that accompanied the exhibition, Pissarro argues, in "The Formation of Crepuscular and Nocturnal Themes in Van Gogh's Early Writings," that Van Gogh had a lifelong love of the night that originated long before he began painting. For more on painting darkness and night, also see *Frederic Remington: The Color of Night* (Washington, D.C.: National Gallery of Art, 2003), by Nancy K. Anderson.

8: Tales from Two Cities

The epigraph from François Jousse comes from Elaine Sciolino's "As the Sun Sets, A Parisian's Masterpiece Comes to Life" (*New York Times*, December 23, 2006, http://www.nytimes.com/2006/12/23/world/europe/23jousse.html?pagewanted=all). An accompanying slideshow shows Jousse in action. See also the Australian Broadcast Corporation's "Foreign Correspondent" feature on Jousse (http://www.abc.net.au/foreign/content/oldcontent/s2464785.htm). The video opens with a short interview with the reigning "Miss Paris," asking if she knows the city's lighting design is mostly the result of a man named François Jousse. She doesn't. By the end of the video, she is saying, "Merci, François Jousse." When I asked Jousse, he laughed and told me that he never met Miss Paris and hadn't known anything about her participation in the video until he saw the finished product.

London overtook Beijing, China, in 1825 with a population of 1,335,000 and held the title of most populous city for a century, when, despite a population well over 7 million, it was passed by New York. Today, the title is generally agreed to be held by Tokyo, Japan, with a population of—depending on how you count—between 13 million and 33 million.

The Scottish writer Robert Louis Stevenson, famous for *Treasure Island* (1883) and "The Strange Case of Dr. Jekyll and Mr. Hyde" (1886), published "A Plea for Gas Lamps" in 1881's *Virginibus Puerisque and Other Papers.* See also his poem "The Lamplighter," which describes a child's desire to "go round at night and light the lamps with you!" It seems apparent that Stevenson was not a big fan of the intensely bright arc lights, those he called "a lamp for a nightmare." He wrote, "Such a light as this should shine only on murders and public crime, or along the corridors of lunatic asylums, a horror to heighten horror." These same lights were soon shining in cities across western Europe and North America.

Charles Dickens's essay "Night Walks," published in 1861, can be found online and in several anthologies, the best of which might be *Night Walks: A Bedside Companion*, compiled by Joyce Carol Oates (Princeton, NJ: The Ontario Review Press, 1982). In her introduction, Oates writes, "There is a nocturnal personality, a nocturnal spirit, distinct from that of daylight and available only in solitude," which certainly characterizes Dickens's fascination with walking through London in the middle of a winter's night.

The book by Sukhdev Sandhu—*Night Haunts: A Journey through the London Night* (London: Verso, 2007)—emerged from a web-based collaboration between writer Sandhu, website designer Ian Budden, and sound artist Scanner. The project can be found online at http://www.nighthaunts.org.uk/ . Starting with the deceptively simple question "Whatever happened to the London night?," Sandhu spends time not only with the bargers on the Thames but many other Londoners working the night shift: police in helicopters, sewer cleaners, mini-cab drivers, graffiti writers, and an urban fox hunter. "I'm keen to reactivate the largely-dormant Victorian and early-twentieth-century genre of the midnight traipse across the metropolis," he writes, but also "to provide . . . its present-day reality."

Virginia Woolf's "Street Haunting: A London Adventure" (1927) can be found online at http://grammar.about.com/od/classicessays/a/strtwoolfessay.htm.

For more on London at night before the advent of electric lighting, see Peter Ackroyd's *London: The Biography* (New York: Anchor, 2003), especially the chapters "Let There Be Light" and "Night in the City." As Ackroyd writes, "It is the presence of the past, or the presence of the dead, which lends the night images of London their peculiar intensity and power. . . . It is an echoic city, filled with shadows, and what better time to manifest itself than at night?"

Originally from San Francisco, David Downie has called Paris home since 1986. His *Paris, Paris: Journey into the City of Light* (New York: Broadway Books, 1990; reprint, 2011) features thirty essays on the "places, people, and phenomena" of the French capital and evocative photographs by Downie's wife, Alison Harris. Together, the two have published several books on France and Italy, with a special attention to food. Because of a degenerative eye disease that makes him highly sensitive to light, Downie is especially a fan of Paris after dark. Though, he told me, he's concerned for the future. "If it gets any brighter," he said, "they're going to have to start calling it the 'city of blinding light.'"

Joachim Schlör's *Nights in the Big City: Paris, Berlin, London, 1840–1930* (London: Reaktion Books, 1998) provides an entertaining "history of the nocturnal city," documenting how over less than a century artificial light radically changed life in Europe's major cities. While detailing the danger of the city at night, Schlör always has "the forgotten beauty of the dark" close at hand, especially as it relates to walking. "The nocturnal walk through the city can appeal to *memory;* it can revive feelings once thought lost, which find no expression in the day; it can awaken a new sense of beauty."

Les Nuits de Paris, or, The Nocturnal Spectator (New York: Random House, 1964) chronicles Bretonne's nighttime adventures in 100 short episodes representing 100 nights of walking through Paris. His conceit is that he had promised a wealthy madame living in the Marais neighborhood he would walk through the city and report back to her. And so we get Bretonne interacting with people on the street—hoodlums, hookers, bakers—but, rather than do as a modern writer might and simply observe, Bretonne is all about interacting with his fellow Parisians, often trying to settle disagreements, rescue young ladies, or slip into and out of parties unnoticed. In her history of walking, Rebecca Solnit claims Bretonne wrote of Paris "in the way many others would later: as a book, a wilderness, and a sort of erogenous zone, or bedroom." The result is entertaining and revealing, perhaps nowhere more than in his account of the violence during the 1789 revolution ("his belly was slashed open, and his head cut off"), which he witnessed firsthand and only narrowly escaped—or so he claims.

Before the more efficient guillotine arrived, the preferred method of execution during the French Revolution was to hang the accused from the nearest lamppost, a common enough event that the verb "lanterner" changed its meaning from "to do nothing" or "to waste one's time" to "to hang someone from the lantern," and the emblematic song of the revolution, "Ah! ça ira," evolved to include the practice:

Ah! ça ira, ca ira, ca ira
Les aristocrates à la lanterne!
Ah! ca ira, ca ira, ca ira
Les aristocrates on les pendra!

Ah! It'll be fine, it'll be fine, it'll be fine
Aristocrats to the lamppost!
Ah! It'll be fine, it'll be fine, it'll be fine
Aristocrats, we'll hang them!

In the most famous instance, on July 22 of 1789, two hated members of the ancien régime, Foulon and Berthier, were hanged in the Place de Grève near the Hôtel de Ville. Bretonne offers a detailed description of the scene, including seeing the blood-soaked corpse of Berthier, and his admission that "I have since heard that the chest was open, and that the heart had been ripped out." Today, while the old lanterns are long gone, a plaque above the street corner marks the spot.

For a fictional—though thoroughly researched—account of Paris in the mid-eighteenth century, see Patrick Suskind's novel *Perfume: the Story of a Murderer* (New York: Knopf, 1986). Suskind gives his main character, Grenouille, an almost supernatural sense of smell and layers his story with details of a time when "there reigned in the cities a stench barely conceivable to us modern men and women"; "the stench was foulest in Paris." Grenouille "worked as long as there was light—eight hours in winter, fourteen, fifteen, sixteen hours in summer," and "had soon so thoroughly smelled out the quarter between St.-Eustache and the Hôtel de Ville that he could find his way around in it by pitch-dark night."

For an excellent account of how artificial light in the streets of seventeenth-century Europe radically changed city life—and other fascinating details, including the way dinnertime shifted several hours later—see Craig Koslofsky's *Evening's Empire: A History of the Night in Early Modern Europe* (New York: Cambridge University Press, 2011). This wave of new light washed over northern Europe quickly, says Koslofsky, reporting that "in 1660, no European city had permanently illuminated its streets, but by 1700 consistent and reliable street lighting had been established in Amsterdam, Paris, Turin, London, and Copenhagen, and across the Holy Roman Empire from Hamburg to Vienna."

E. Roger Ekirch's *At Day's Close: Night in Times Past* (New York: Norton, 2005) offers readers an entertaining history of pre-industrial night in western Europe and eastern North America. With exhaustive research and an often bemused tone, Ekirch presents page after page of eye-opening facts and stories about the Western world before the electric light. The book is perhaps best known for Ekirch's discovery of the segmented sleep patterns followed by pre-industrial peoples, but his discussion of "the way of life people fashioned after dark in the face of both real and supernatural perils" covers everything from the catastrophic danger of fire to the humor of mistaken identities in the darkened bedroom.

During the 1830 revolution, lantern smashing became a street-fighting tactic used against government forces, and it later found its way into Victor Hugo's novel *Les Miserables* (1862), where the practice turned the city's center into "a huge patch of darkness...a black gulf." In a short chapter titled "The Street Urchin an Enemy of Light," the hero, Valjean, encounters a street urchin bent on destroying every lantern he can find. After "the lamp-lighter came as usual to light the lantern," Valjean sees "by the light of the lanterns" the boy's face. As the two are talking, the boy picks up a stone and destroys the lantern with a well-placed shot, shouting, "That's right, old street, put on your night-cap." When Valjean offers the boy a coin for food, the boy rejects it, saying, "Bourgeois, I prefer to smash lanterns." When the boy finally flutters away "like an escaped bird," plunging "back into the gloom as though he made a hole it," Valjean wonders if the whole encounter had really taken place. But then, in the distance, "a startling shiver of glass" and "the magnificent crash of a lantern rattling down on the pavement."

The photographs of Paris at night in the 1930s by the Hungarian photographer Brassaï are breathtaking in their depiction of the city's new electric streetlights pushing back against the eternal weight of old night. Of Brassaï's 1933 masterpiece *Paris by Night* (New York: Pantheon, reprint, 1987), originally published in English as *Paris After Dark*, Joachim Schlör writes, "the book expressed a certain feeling of the times, that something was coming to an end: a long history of the Parisian night." In the book, Brassaï's photographs are accompanied by the words of Paul Morand ("Moving on, I hear the grisly sound of pigs being sawn in two"), whose brief introduction argues, "night is not the negative of day... as on the photographic plate... another picture altogether emerges at nightfall."

In addition to rejecting the idea of having Notre Dame's rose window lit from within, the cathedral priests also denied an idea from lighting designer Roger Narboni to have a wave of light pulse the length of the cathedral at the top of each hour like the tone of a bell. As Narboni told me, "From the beginning I wanted to have a wave of light, piloted by computer, that every hour would start from the back and move slowly through to the main façade. I called it the luminous bell of the cathedral, because there are big, big bells in the towers but they don't ring them anymore on the hour. It would last for two, three, or four minutes and that's it. But again, the church didn't want that. They said we don't want any dynamic light on the cathedral; this is a cathedral, it's not a disco. And I said this has nothing to do with a disco, it's very slow. But they never understood the poetry of it. So, in the end everyone said forget about your wave, forget about your bell."

For more on Gustave Eiffel's famous tower see Jill Jonnes's *Eiffel's Tower: The Thrilling Story Behind Paris's Beloved Monument and the Extraordinary World's Fair That Introduced It* (New York: Penguin, 2009). Jonnes shares the bewildering fact that the tower was only built for the Paris Fair of 1889 with the understanding that it would be torn down soon after (it was saved only through Eiffel's ingenuity in rallying support for his creation). She also explains the challenges Eiffel faced in building the iconic tower (such as engineering all four legs to reach the first platform exactly) and tells of what are to our ears today the ridiculous attacks against the tower (that it could not stand the power of the wind, that it would "become a dangerous magnet, drawing the nails from surrounding Parisian buildings"). In an era when the tower has become all but synonymous with Paris, the story of its creation and survival is a fascinating and enjoyable read.

7: Light That Blinds, Fear That Enlightens

The quote from Annie Dillard can be found in "Seeing," from her book *Pilgrim at Tinker Creek* (New York: Harper & Row, 1974), which won the 1974 Pulitzer Prize for General Nonfiction.

The "darkness" I experienced on the suburban golf course is explained by the finding that clouds amplify city lights by ten times. See the article by Christopher Kyba and others titled "Cloud Coverage Acts as an Amplifier for Ecological Light Pollution in Urban Ecosystems," part of the Verlust der Nacht (Loss of the Night), a federally funded study in Germany (www.verlust-dernacht.de).

Liability fears are one of the key reasons we use so much light in American society, especially when it comes to places such as parking lots and college campuses. No one wants to be sued for a crime or accident that occurred where there wasn't "enough" light. Yet in reality the established law when it comes to this subject is virtually nonexistent. At the very least, as one lawyer told me, "There is no law requiring landowners to illuminate their property excessively, and there is no law requiring landowners to illuminate the heavens."

The quote about our choice being either a view of a good night sky ("less street lighting") or fearing for our lives ("attacked by a violent predator") comes from a 2006 position paper titled "Dark Sky Ordinances: How to Separate the Light from the Darkness," by David B. Kopel and Michael Loatman (http://www.scribd.com/doc/29812975/Dark-Sky-Ordinances-How-to-Separate-the-Light-from-the-Darkness) of the Independence Institute, a conservative think tank based in Colorado. The paper uses assertions both unfortunate ("dark sky ordinances mainly benefit casual urban stargazers") and overstated to the point of being false ("research shows that improved street lighting reduces crime by 20 percent") to rail against attempts to control overuse of light. The claim that "the overall reduction in crime after improved lighting was 20%" comes from a 2002 report titled "Improved Street Lighting and Crime Prevention," by David Farrington and Brandon Welsh.

Part of the difficulty in any discussion about lighting and safety are the terms we use. What exactly do we mean by "better lighting," "well-lighted," and "improved lighting," for example? Were we to understand these terms to mean lighting that was fully shielded and sensitively designed, that would be one thing. Unfortunately, when it comes to traditional understanding of light and safety, usually these terms simply mean bright or brighter.

The information on lighting efficiency and per capita electricity consumption in the UK comes from "Seven Centuries of Energy Services: The Price and Use of Light in the United Kingdom (1300–2000)," by R. Fouquet and P. Pearson (*The Energy Journal* 27 [2006]: 139–77).

The British Astronomical Association's Campaign for Dark Skies (CfDS) offers a wealth of information on the benefits of darkness and the perils of bad lighting (http://www.britastro.org/dark-skies). The CfDS has as its goal "to preserve and restore the beauty of the night sky by campaigning against excessive, inefficient, and irresponsible lighting that shines where it is *not* wanted nor needed."

The passage from Ralph Waldo Emerson comes from *Nature*, published in 1836. The essay has no particular connection to the night sky much beyond this early passage—Emerson's interest in the stars is primarily symbolic. In the next line he writes, "The stars awaken a certain reverence, because though always present, they are inaccessible; but all natural objects make a kindred impression, when the mind is open to their influence." One wonders what Emerson might say about the stars and their ability to inspire reverence were he alive to witness their loss from most of our skies.

For a study of gas station canopy lighting that shows "how a gas station can provide satisfactory light levels... while reducing glare and light trespass," see http://www.lrc.rpi.edu/programs/transportation/pdf/lightPollution/canopy.pdf.

The terms "lux" (as in Roger Narboni starting with "400 lux on the fish") and "footcandles" are the units ("lux" being metric) for the amount of light present at a given surface, or illuminance. We cannot really "see" illuminance, but light meters can measure it. Light we perceive either from surface brightness (reflected light) or a light source is luminance. Neither illuminance nor luminance is to be confused with "lumens," the measure of the amount of light given off by a source of light.

Of life as a lighting designer in a society obsessed with equating bright lights and safety, Narboni told me, "And now we have this crazy politics about security in the cities that has totally changed the way we work. We have to fight a lot against the politicians who want to light everything with a high level of light because they think this will solve everything about vandalism and delinquents and everything and this is a crazy idea. So no one wants shadows or darkness—it's a fight that we lose almost every time. It's terrible."

The Campaign for Dark Skies has an entire page devoted to towns that have chosen to turn down or turn off unnecessary lighting at night: www.britastro.org/dark-skies/lightsoffresponse.html. Information on the city of Bristol can be found at www.thisisbristol.co.uk/Burglars-afraid-dark-Crime-falls-Bristol-street/story-13952633-detail/story.html, and on Rockford, Illinois, at www.npr.org/2011/11/08/142145523/rockford-ill-shuts-off-streetlights-to-save-money. It's hard to find a better example of a city website that explains the benefits of reducing the amount of light used at night than that of Santa Rosa, CA (http://ci.santa-rosa.ca.us/departments/publicworks/streetlightreduction/Pages/default.aspx).

One study reporting criminals' perspectives on lighting can be found at http://www.policypointers.org/Page/View/1238. The final report from the Chicago Alley Lighting Project can be found at http://www.icjia.state.il.us/public/pdf/ResearchReports/Chicago%20Alley%20Lighting%20Project.pdf. The International Dark-Sky Association website has a wealth of position papers and links to studies related to lighting and safety/security (darksky.org). Another useful source is the Royal Astronomy Society of Canada, Calgary Centre (http://calgary.rasc.ca/lp/index.html).

Find Dr. Barry Clark's invaluable reviews of the available research on "outdoor lighting and crime" at http://asv.org.au/light-pollution.php. For anyone interested in this issue, Clark's work is simply required reading. Of the position paper from the Independence Institute, Clark told me, "Its chorus of half-truths takes up so much of the score that it is almost a continuous perversion of scientific method. The authors might now turn their hands to justifying the continuing use of asbestos... and promoting smoking for teenagers."

The quotes on fear of the dark from E. Roger Ekirch come from the first few pages of *At Day's Close.* It's significant that in this 350-page history Ekirch begins this way, an acknowledgment that our fear of the dark—whether subconscious or not—absolutely colors our relationship with night. A 2010 History Channel documentary titled *Afraid of the Dark* featured Ekirch extensively as it detailed the reasons for our fear, including "belief in ghosts," "the supernatural," "Satan," "wild animals," and "deadly terrain." Sitting bathed in TV glow in our well-lit houses in our well-lit suburbs or cities, we might chuckle at these primitive fears, but place us in a truly dark night without electric light and we might feel differently.

For another excellent take on fear of the dark, see "The Dark at the Top of the Stairs," from A. Alvarez's *Night: Night Life, Night Language, Sleep, and Dreams* (New York: Norton, 1995). Alvarez writes, "Fear of the dark is essentially unspecific; like darkness itself, it is formless, engulfing, full of menace, full of death." Because of this, "in horror movies, no matter how brilliant the special effects, the moment when the monster is finally revealed is invariably a disappointment."

Katie Roiphe's comments about the blue-light system on college campuses can be found in *The Morning After: Sex, Fear, and Feminism* (Boston: Back Bay, 1993). In fewer than two decades, these blue lights atop silver poles have swept across the landscape of college and university campuses in the United States. Rare is the institution that has not spent many thousands of dollars to purchase, install, and maintain the lights. Rarer still is any research proving their effectiveness at actually making anyone safer rather than simply making us "feel" safer.

The report on "The Sexual Victimization of College Women" can be found at https://www.ncjrs.gov/pdffiles1/nij/182369.pdf. The article by Jennifer K. Wesely and Emily Gaarder titled "The Gendered 'Nature' of the

Urban Outdoors: Women Negotiating Fear of Violence" comes from *Gender and Society* 18, no. 5 (October 2004).

It's not until page 233 of her 291-page book *Wanderlust: A History of Walking* (New York: Penguin, 2000) that Rebecca Solnit writes, "Throughout the history of walking I have been tracing, the principal figures... have been men, and it is time to look at why women were not out walking too." She then enters perhaps the most compelling section of her fascinating book, explaining how she was nineteen before she "first felt the full force of this lack of freedom" that being a woman entailed. "I was advised to stay indoors at night," she writes, and the messages she received asserted that it was her responsibility "to control my own and men's behavior rather than society's to enforce my freedom" to walk after dark.

Brianna Denison, a nineteen-year-old student visiting a friend at the University of Nevada in Reno, was abducted in January of 2008. Her body was found in mid-February. Her assailant was caught and sentenced to death. Twenty-two-year-old Eve Marie Carson was murdered March 5, 2008, in Chapel Hill, North Carolina. Her twenty-one- and seventeen-year-old assailants were also caught and sentenced to life without parole.

Motor vehicle deaths in the United States reached their zenith in 1972, when more than 54,000 Americans were killed. Due especially to improvements in automobile safety features, motor vehicle deaths in this country have been falling gradually since. In 2010, with the country's population having increased by nearly 100 million from four decades before, the death toll was 32,708.

According to author Barry Lopez, our identification of evil with nocturnal animals comes from deep within us. In his *Of Wolves and Men* (New York: Scribner, 1978), he writes, "Killing wolves has to do with murder. Historically, the most visible motive, and the one that best explains the excess of killing, is a type of fear: theriophobia. Fear of the beast. Fear of the beast as an irrational, violent, insatiable creature.... In its headiest manifestations theriophobia is projected onto a single animal, the animal becomes a scapegoat and it is annihilated." It is estimated that, from the time of the arrival of the first European colonists in North America, the wolf population fell from more than 250,000 to fewer than 1,000, its range reduced to perhaps 3 percent of its historic limits. Other sources say that one to two million were killed in the latter half of the nineteenth century alone. As of 2012, the wolf population in the Lower 48 has climbed back above 5,000.

Among Ken Lamberton's fine books are *Beyond Desert Walls: Essays from Prison* (2005), *Dry River: Stories of Life, Death, and Redemption on the Santa Cruz* (2011), and *Wilderness and Razor Wire* (1999), which won the 2002 John Burroughs Award for nature writing—all published by the University of Arizona Press in Tucson. His essay "Night Time" can be found in Paul Bogard's *Let There Be Night: Testimony on Behalf of the Dark* (Reno: University of Nevada Press, 2008).

The subject of prison lighting is one that not many people think about—except perhaps for the prisoners themselves and those who work in prisons, all of whom spend long hours in poorly lit environments. Michaele Wynn-Jones has worked for more than fifteen years on this problem and claims, among other consequences, that poor prison lighting—both the type of lighting and its near-constant presence—has significantly contributed to incidences of depression and suicide among those who work and live in prison. "Imagine living up to twenty-three hours a day in a confined space the size of the average bathroom...under a humming fluorescent tube that is your only available light source," she writes ("Life under Fluorescent Light Is Harming Prisoners and Staff Alike," in the *Guardian*, September 26, 2002).

On a more positive note, some prisons in California have begun to install "eco-friendly" lighting in an effort to save energy, a significant cost consideration in facilities where lighting systems run twenty-four hours a day, every day of the year.

Aldo Leopold's story of Escudilla Mountain in southern Arizona can be found in *A Sand County Almanac* (New York: Oxford University Press, 1949). In 1984, Congress declared 5,200 acres of the mountain and surrounding Apache National Forest as official wilderness. September and October, when the mountain's many aspen trees are changing, is a wonderful time to visit. There are no grizzly bears to worry about, or to miss.

6: Body, Sleep, and Dreams

The World Health Organization's International Agency for Research on Cancer (IARC) listed shift work involving circadian disruption a "probable human carcinogen" in 2007. For a summary of what might have led it to do so, see "Considerations of Circadian Impact for Defining 'Shift Work' in Cancer Studies: IARC Working Group Report" (*Occupational Environmental Medicine* 68 [2011]: 154–62).

In 2009, the American Medical Association voiced its unanimous support for "light pollution control efforts and glare reduction for both public safety and energy safety," declaring, in part: "Whereas, Our AMA has long advocated for policies that are scientifically sound and that positively influence public health policy; and...Whereas, Light trespass has been implicated in disruption of the human circadian rhythm, and strongly suspected as an etiology of suppressed melatonin production, depressed immune systems, and increase in cancer rates....Therefore be it Resolved, That our AMA support light pollution reduction efforts and glare reduction efforts at both the national and state levels." In 2012, the AMA went further, adopting new policy "recognizing that exposure to excessive light at night can disrupt sleep, exacerbate sleep disorders and cause unsafe driving conditions."

An excellent summary of the connections between light at night and health can be found in "Missing the Dark: Health Effects of Light Pollution" by Ron Chepesiuk (*Environmental Health Perspectives* 117 [2009]: A20–A27). The Campaign for Dark Skies has a helpful page on the connections between light at night and human health: http://www.britastro.org/dark-skies/health.html.

The comments from Eva Schernhammer about the risks from working at night come from "Light at Night and Health: The Perils of Rotating Shift Work" (*Occupational and Environmental Medicine*, October 4, 2010). The comments from Chuck the locomotive engineer come from "Working the Graveyard Shift, Fighting the Sandman," from NPR's *Talk of the Nation*, April 26, 2011.

For information on who works at night, including the high percentage of African Americans and the number of women who report higher work-family conflict, see the Bureau of Labor Statistics as reported in "Opportunities for Policy Leadership on Shift Work," from the Sloan Work and Family Research Network (http://workfamily.sas.upenn.edu/sites/workfamily.sas.upenn.edu/files/imported/pdfs/policy_makers6.pdf).

For a highly readable article detailing the potential connections between light at night and cancer, see Richard G. Stevens's "Light-at-Night, Circadian Disruption and Breast Cancer: Assessment of Existing Evidence" (*International Journal of Epidemiology* 38 [2009]: 963–70). At the very least, Stevens concludes, "increasing numbers of people must do shift work in modern societies, and few people will give up electric lighting at home. An understanding of what particular characteristics of wavelength, intensity, timing, and duration most disrupt circadian rhythms would permit a minimization of any potential health risks."

The possible connections between breast cancer and the blue light of computers and televisions are detailed in "The Light-Cancer Connection" by Catherine Guthrie (*Prevention* 58, no. 1 [January 2006]).

Read more from Steven Lockley about the consequences of not getting enough sleep in *Sleep: A Very Short Introduction* (New York: Oxford University Press, 2012), cowritten with Russell G. Foster, and in the International Dark-Sky Association's *Fighting Light Pollution: Smart Lighting Solutions for Individuals and Communities* (2012).

Statistics on how many of us are not getting enough sleep are readily available. For example, see the Centers for Disease Control's "Insufficient Sleep Is a Public Health Epidemic" (http://www.cdc.gov/features/dsSleep). What isn't readily available are statistics on the relation of long light to short sleep, and on the potentially huge cost of electric lighting to our health and economy.

Many of the night-shift nurses whom I asked said that in order to sleep during the day they must use blackout curtains on their bedroom windows. As Michelle in St. Paul told me, "Personally, I would have trouble sleeping in a bright room," a seemingly obvious statement until one thinks about how many of us try to sleep at night with bright sources of light coming through our windows. The trade-off, at least for nurses in the United States working the night shift, almost always includes such perks as a few extra dollars per hour in salary, extra vacation hours, and free parking.

In a valiant effort to highlight the decline of the afternoon siesta tradition, Spain's National Association of Friends of the Siesta organized, in October 2010, the country's first siesta championship, with contestants ranked for loudest snore, most original sleep positions, and duration of sleep. The winner took home a prize of 1,000 euros.

The connection between light at night and obesity emerged through a study on mice led by Laura Fonken of Ohio State University. Fonken and her team of researchers divided mice into three groups: one group lived with a natural light-dark cycle, a second group endured constant light, and the third group had the darkness of their light-dark cycle replaced with a dim glow. The researchers found that the mice in the second and third groups gained almost 50 percent more weight than the mice in the first group. The members of the two groups also gained more fat than those in the first group and showed a reduced tolerance for glucose.

Vaughn McCall tells his patients to divorce themselves from the clock. "A common problem I see in insomniacs is that even if they have the rest of the room pitch-black dark, they have a clock," he told me. "And the clock becomes their master. They become slave to it and they start fretting over, 'My God, I've been awake ten minutes. What if I'm awake fifteen minutes? How long has it been now? Oh, soon it's going to be twenty minutes.' And I tell them, 'Actually, you will be more comfortable with being awake if you get rid of the clock or at least just turn the thing around. There's nothing that the clock is going to tell you in the middle of the night that's going to make you feel any better about yourself.'"

Rubin Naiman's book *Healing Night: The Science and Spirit of Sleeping, Dreaming, and Awakening* (Minneapolis: Syren Book Co., 2006) deserves a wider audience than it has received. Arguing that traditional sleep medicine "makes absolutely no allowance for the spiritual dimensions of night, sleep, or dreams," and decrying our society's "undeclared war against dusk and darkness," Naiman makes a persuasive case for a more holistic approach to our experience of night, sleep, and darkness. We are still afraid of the dark, he argues, and "our disturbed relationship with night is ultimately rooted in our discomfort with and denial of the dark side of our own selves."

Naiman told me that one of his favorite books is Dr. Seuss's *Sleep Book* (New York: Random House, 1962), which includes a character named the Chippendale Mupp, who has an unbelievably long tail. "When it goes to bed at night, it gathers the tail—it takes some time—and finally, after a while, when it gets the end of the tail, it chomps down on its own tail really hard, and then it goes to sleep," Naiman explained. "Because the tail is so long, it takes precisely eight hours for that nip, as he calls it, that pain impulse to come back to the Mupp's brain. I'd read this many times and I finally realized, 'Oh, my God! He's teaching kids the truth about an alarm clock!' "

Here's the full passage of Thoreau fishing at night, from *Walden:* "Sometimes, after staying in a village parlor till the family had all retired, I have returned to the woods, and, partly with a view to the next day's dinner, spent the hours of midnight fishing from a boat by moonlight, serenaded by owls and foxes, and hearing, from time to time, the creaking note of some unknown bird close at hand. These experiences were very memorable and valuable to me,—anchored in forty feet of water, and twenty or thirty rods from the shore, surrounded sometimes by thousands of small perch and shiners, dimpling the surface with their tails in the moonlight, and communicating by a long flaxen line with mysterious nocturnal fishes which had their dwelling forty feet below, or something dragging sixty feet of line about the pond as I drifted in the gentle night breeze, now and then feeling a slight vibration along it, indicative of some life prowling about its extremity, of dull uncertain blundering purpose there, and slow to make up its mind. At length you slowly raise, pulling hand over hand, some horned pout squeaking and squirming to the upper air. It was very queer, especially on dark nights, when your thoughts had wandered to vast and cosmogonal themes in other spheres, to feel this faint jerk, which came to interrupt your dreams and link you to Nature again. It seemed as if I might next cast my line upward into the air, as well as downward into this element, which was scarcely more dense. Thus I caught two fishes as it were with one hook."

5: The Ecology of Darkness

Anyone who lived two years, two months, and two days in the Massachusetts woods during the mid-nineteenth century would have known well the primitive darkness of natural night. Reading *Walden* with this in mind, it is clear that for Henry David Thoreau this was the case. First published in 1854, by Ticknor and Fields in Boston, with the seldom-used full title *Walden; or, Life in the Woods,* the book features chapters titled "Sounds," "Solitude," and "The Village," in which we find direct reference to darkness John Bortle would have ranked at 1 (at the pond) or 2 (in Concord). In a later essay, "Night and Moonlight," Thoreau's connection to the dark is clear. How ironic, then, that Walden Pond State Reservation closes at sunset and Thoreau's cabin remains unvisited every night of the year. Of course, that the site remains preserved at all is worth celebrating—as recently as 1990 it took the rock-and-roll star Don Henley creating the Walden Woods Project to keep development from radically altering the area. While Thoreau's cabin has long been gone—as has the darkness he knew—the nearby Concord Museum has an excellent collection of his artifacts, and the pond remains a wonderful place to visit and to remember a writer whose reflections on our way of life seem to grow more applicable by the year.

Thoreau's "Night and Moonlight" was published in the *Atlantic Monthly Magazine* in November of 1863, some six months after his death. "Chancing to take a memorable walk by moonlight some years ago," he begins, "I resolved to take more such walks, and make acquaintance with another side of Nature. I have done so." He seems to have understood late in his very short life (he was only forty-four when he died) how rich the world of night and moonlight could be for his thinking and writing. He asks, "What if one moon has come and gone, with its world of poetry, its weird teachings, its oracular suggestions,—so divine a creature freighted with hints for me, and I have not used her,—one moon gone by unnoticed?" One wonders, as with Emerson, what he would say about our light-washed world today.

Hearing the Walden frogs brings to mind "From Silent Spring to Silent Night," the compelling work put forward by UC–Berkeley's Dr. Tyrone Hayes linking the use of the pesticide atrazine with the decline in frog numbers, and what this loss means, among other things, for the soundscapes of our nights.

For a detailed discussion of the distinction between "wilderness" and "wildness," see William Cronon's "The Trouble with Wilderness; or, Getting Back to the Wrong Nature," from his book *Uncommon Ground: Rethinking the Human Place in Nature* (New York: Norton, 1996).

In addition to gathering the scattered research on wildlife and darkness, one of the features that makes *Ecological Consequences of Artificial Night Lighting* (Washington, D.C.: Island Press, 2006), edited by Catherine Rich and Travis Longcore, unique is its attempt to combine scientific work with creative literature. The book's epigraph comes from Thoreau's "Night and Moonlight," and each section begins with a short selection of creative work by such authors as Bernd Heinrich and Carl Safina. As with so many issues related to the natural world, the information gathered by scientists about the impact of artificial lighting on wildlife (let alone on humans) will only be as powerful as the stories used to present it. While Rachel Carson's *Silent Spring*, for example, was full of scientific research, her book would not have been as powerful without her metaphoric title or her opening "Fable for Tomorrow." Rich and Longcore deserve credit for their efforts to bring attention to the "wasted, ecologically disruptive light" now filling our nights, and for their argument that this light is "itself the end product of extractive and consumptive processes that are themselves environmentally damaging."

Funded by Germany's Federal Ministry of Education and Research, the ongoing study called Verlust der Nacht is the most promising work addressing ecological light pollution. Researchers involved note that while attention to the economic costs of lighting and light pollution are important, knowledge is urgently needed for "light pollution policies that go beyond energy efficiency to include human well-being (and) the structure and functioning of ecosystems." In an early paper titled "The Dark Side of Light," participants in Verlust der Nacht warn that "unless managing darkness becomes an integral part of future conservation and lighting policies, modern society may run into a global self-experiment with unpredictable outcomes."

Perhaps the most comprehensive book on ecology and night is *Nightwatch: The Natural World from Dusk to Dawn* (London: Roxby & Lindsey Press, 1983). Featuring text by seven different writers and stunning photographs by Jane Burton and Kim Taylor, *Nightwatch* offers an exhaustive accounting of the value of darkness for the wild world. From sleep and tides and biological clocks to chapters on woodlands and freshwater and the sea at night, this book is a prescient look at a world under threat from human overuse of artificial light. That there has been no comparable book in the thirty years since, even while light pollution has grown by leaps and bounds, says much about our inattention to the wild world at night and to the negative effects of artificial light on that world.

One of the most dramatic images from Verlyn Klinkenborg's wonderful *National Geographic* article "Our Vanishing Night" (November 2008) is the back-to-back photographs of Los Angeles taken from Mt. Wilson, first in 1908 and then one hundred years later. In the 1908 photograph, the city of three hundred fifty thousand sits surrounded by dark countryside, while in the image from 2008 the city of five million fills the entire frame with a shimmering swath of electric light. As one result of this change, the observatory at Mt. Wilson was rendered useless for optical astronomy and essentially abandoned by its previous owner, the Carnegie Institution, to the Mount Wilson Observatory Association for just $1.

Though obviously not the result solely of incidents occurring at night, the roadkill numbers in the United States are staggering: at least one million vertebrates per day (birds, mammals, reptiles, amphibians). Even in locations ostensibly designed to be safe havens for wildlife, there seems to be no escape. From 1989 to 2003 in Yellowstone National Park, for example, some 1,559 animals were killed by cars, including 556 elk, 192 bison, 135 coyotes, 112 moose, 24 antelope, and 3 bobcats (U.S. Department of Transportation). The good news is that thoughtfully designed fences, culverts, crosswalks, overpasses, etc., can significantly reduce vehicle-animal collisions.

For more information on Civil Twilight and their idea for "lunar-resonant streetlights," see http://www.metropolismag.com/story/20070518/lunar-light. One member of Civil Twilight, Christina Seeley, is a wonderful photographer. Her series titled *lux* documents sky glow in the United States, western Europe, and Japan. See her work at www.christinaseely.com.

Saying that we "simmer in our own electronic bouillabaisse" of light, James Attlee sets out to reestablish "a lost connection with the moon" in

Nocturne: A Journey in Search of Moonlight (Chicago: University of Chicago Press, 2011).

Few have done as much as author and illustrator John Himmelman to draw our attention to the beauty and value of moths, crickets, and other insects that do so much to make our nights (and our world) come alive. *Cricket Radio: Tuning in the Night-Singing Insects* (Cambridge, MA: Harvard University Press, 2011) is his latest contribution, and *Discovering Moths: Nighttime Jewels in Your Own Backyard* (Camden, ME: Down East Books, 2002) is especially rich. His words about the luna moth appear on pages 81–82.

Bat Conservation International (BCI), started by Merlin Tuttle, continues to work on behalf of bats around the world. Their website (batcon.org) has a wealth of information on the importance of bats and the threats they face. Among this information is everything you need to know about the bats living under Congress Avenue Bridge in Austin, Texas, including emergence times from March through October. Bat tourism is big business in Austin, with millions of dollars generated by an estimated one hundred thousand people every year who watch the bats spiral out from under the bridge and swirl off into the surrounding countryside. BatCon members are invited to witness the even more dramatic emergence of millions of bats at the BCI-owned Bracken Bat Cave, on the edge of San Antonio. Though somewhat dated, the BCI-produced DVD *The Secret World of Bats* offers forty-eight minutes' worth of video designed to debunk myths and encourage admiration. The slow-motion video of bats pollinating cactus flowers is especially impressive.

When I asked Merlin Tuttle about bats attacking humans, he told me that even rabid bats are rarely aggressive. "In more than fifty years of studying bats, often in caves with millions, I have never been attacked, nor have any of the millions of tourists who have closely observed bats at the Congress Avenue Bridge in Austin over the past thirty years. The odds of being harmed by a bat," he emphasized, "are exceedingly remote for anyone who simply doesn't attempt to handle them."

The study showing the economic benefits brought to humans by bats, "Economic Importance of Bats in Agriculture," can be found at http://www.sciencemag.org/content/332/6025/41. At the other end of the spectrum from the estimated $54 billion benefit bats provide us is the very small amount we spend to protect them (only $2.4 million in 2010).

Knowing that communication towers continue to rise in ever more remote locations, and wanting to "provide a scientific basis for regulation of tower construction and operation," Travis Longcore, Catherine Rich, and Sidney Gauthreaux have found that steady-burning lights combined with ever-higher structures and heavy use of guy-wires to support the towers prove a lethal combination for birds flying at night. "The towers that are killing big numbers of birds have steady-burning lights on them," Longcore says, because the steady-burning lights hold the birds' attention, drawing them off course and "trapping" them. The good news is that lights that alternate from on to off "release" the birds from their pull, and as a result "you can reduce mortality by 60 to 80 percent just by switching the lighting type." Longcore, Rich, and Gauthreaux write that "avian mortality would be reduced by restricting the height of towers, avoiding guy-wires, using only red or white strobe-type lights as obstruction lighting, and avoiding ridgelines for tower sites."

Find out more about FLAP and its efforts on behalf of migrating birds at www.flap.org. One recent victory for FLAP and others concerned with birds is the adoption by the U.S. Green Building Council's LEED (Leadership in Energy and Environmental Design) of the "bird collision issue" into its rating system. "Pilot Credit 55: Bird Collision Prevention" will require making a building visible as a physical barrier to birds during the day and eliminating light trespass at night. Some one billion birds die every year in the United States due to collision with human-made objects, the vast majority of them flying into glass buildings.

David Gessner's essay "Trespassing on Night" appears in *Let There Be Night: Testimony on Behalf of the Dark* (Reno: University of Nevada Press, 2008). Find out more about his work at davidgessner.com.

Henry Beston's *The Outermost House: A Year of Life on the Great Beach of Cape Cod*, first published in 1928, has been available since 1992 from Henry Holt, New York. Of his chapter "Night on the Great Beach," Beston wrote to his future wife, Elizabeth Coatsworth, during his time at the house he designed and named the Fo'castle, "My last chapter was on 'Night on the Great Beach' and I let myself go, for there's a soupçon of the noctambule in me; or—the noctamorist, I love night."

4: Know Darkness

Rainer Maria Rilke's "You, Darkness" comes from his *Book of Hours (Das Stundenbuch)*, written from 1899 to 1903. For a fascinating look at this poem translated six different ways, see http://www.beyond-the-pale.co.uk/rilke.htm. The themes of darkness and night, literal and metaphorical, flow through the German poet's work. Two especially relevant poems are both titled "Night," the first from 1906 ("The lamps keep swaying, fully unaware: / is our light *lying*?/ Is night the only reality / that has endured through thousands of years?") and the next from 1924:

Night, full of newly created stars that leave
trails of fire streaming from their seams
as they soar in inaudible adventure
through interstellar space:

how, overshadowed by your all-embracing vastness,
I appear minute!——
Yet, being one with the ever more darkening earth,
I dare to be in you.

For more on Chaco Canyon (Chaco Culture National Historical Park), see http://www.nps.gov/chcu/index.htm. There is no shortage of books featuring Chaco Canyon. See especially Craig Childs's *House of Rain: Tracking a Vanished Civilization Across the American Southwest* (New York: Little, Brown, 2007), and Anna Sofaer's *Chaco Astronomy: An Ancient American Cosmology* (Santa Fe: Ocean Tree Books, 2007). Her PBS documentaries *The Sun Dagger* (1982) and *The Mystery of Chaco Canyon* (2000), both narrated by Robert Redford, did much to hype the intrigue surrounding the canyon.

The kivas found at Chaco Canyon are thought to be forerunners of the subterranean kivas used by modern Pueblo peoples for religious rites and rituals.

Published in 1933, Jun'ichiro Tanizaki's *In Praise of Shadows* is still readily available (Sedgwick, ME: Leete's Island Books, 1977). While the genre of elegy is prominent in nature writing (or environmental writing)—so much so that poet Alison Deming asked in a 2000 essay that we might move "beyond elegy" ("Getting Beyond Elegy," *Georgia Review* 54, no. 2 [Summer 2000]: 259–71), I know of no other works (save perhaps Beston's *Outermost House*) that are so directly elegiac in their attention to darkness. Writing at roughly the same time as Beston, though half a world away, Tanizaki saw a bright future and mourned what was being lost. "So benumbed are we nowadays by electric lights that we have become utterly insensitive to the evils of excessive illumination," he argued. "I have written all this because I have thought that...I would call back at least for literature this world of shadows we are losing."

While I knew Joseph Bruchac's name from other books, I did not know of his interest in the night until I found *Keepers of the Night: Native American Stories and Nocturnal Activities for Children* (Golden, CO: Fulcrum, 1994). For more on Native American views of the night sky, see *They Dance in the Sky: Native American Star Myths,* by Jean Guard Monroe and Ray A. Williamson (New York: Houghton Mifflin, 1987). For more on specifically Southwest Native American views, see *Sharing the Skies: Navajo Astronomy,* by Nancy C. Maryboy and David Begay (Tucson: Rio Nuevo, 2010).

Eric Wilson's books include *Against Happiness* (New York: Farrar, Straus & Giroux, 2008), *The Mercy of Eternity: A Memoir of Depression and Grace* (Evanston, IL: Northwestern University Press, 2010), and, most recently, *Everyone Loves a Good Train Wreck: Why We Can't Look Away* (New York: Farrar, Straus & Giroux, 2012). Regarding his titles, I am reminded of his description of Jesus Christ as "a man of sorrows whose melancholy suffering was inseparable from his illumination." Find the Carolina Chocolate Drops at www.carolinachocolatedrops.com.

Anyone wishing to understand the absolute anguish that is depression can do no better than read William Styron's *Darkness Visible: A Memoir of Madness* (New York: Vintage, 1990), which emerged from his spellbinding essay for *Vanity Fair:* http://www.vanityfair.com/magazine/archive/1989/12/styron198912. Whether a sufferer oneself wishing for empathy from another, or a family member, friend, or stranger wishing to understand, Styron's book offers a beautifully written first-hand account of a horrific experience.

James Galvin's *The Meadow* (New York: Holt, 1993) tells a hundred-year history of his neighbors' lives on the Wyoming/Colorado border. A nonfiction work that reads like a novel, *The Meadow* is beautifully, truthfully, imaginatively written, such that if a character says he can hear the stars on the coldest winter nights, the reader believes in and wonders at the sound.

The sounds of night, the quiet of night, the noise of night—so much of our experience of darkness has to do with senses other than sight, and especially with what we hear. In his quietly heartbreaking book *The Great Animal Orchestra: Finding the Origins of Music in the World's Wild Places* (New York: Little, Brown, 2012), Bernie Krause tells of his lifelong quest to record the wild sounds of the earth. "As a seasoned listener, I especially love the sounds produced by creatures that have evolved to vocalize at night," he writes. "The nighttime imparts the sense of a resplendent echoey theater—a beneficial effect for nocturnal terrestrial creatures whose voices need to carry over great distances."

The arguments made by Richard Louv in his best-selling *Last Child in the Woods: Saving Our Children from Nature-Deficit Disorder* (Chapel Hill, NC: Algonquin, 2005) could easily be applied to our children's experience of night and darkness. It bears repeating that estimates suggest that eight of ten American children born today will never live where they can see the Milky Way. A "deficit" signifies our not having enough of something. This is exactly what we allow our children (and ourselves) of darkness: too small an experience, not nearly enough.

In a letter from 1903 collected with others and published in 1934 after his death, Rilke urged his reader in *Letters to a Young Poet* to "love the questions," and so urged millions more who have found inspiration and guidance in his words. "I want to beg you, as much as I can, dear sir, to be patient toward all that is unsolved in your heart and to try to love the questions themselves," he wrote, "like locked rooms and like books that are written in a very foreign tongue."

For an excellent translation of St. John of the Cross's *Dark Night of the Soul*, see the version by Mirabai Starr (New York: Riverhead, 2002). For two contemporary works inspired by St. John's Dark Night, see Gerald May's *The*

Dark Night of the Soul: A Psychiatrist Explores the Connection Between Darkness and Spiritual Growth (New York: HarperCollins, 2004) and Thomas Moore's *Dark Nights of the Soul: A Guide to Finding Your Way Through Life's Ordeals* (New York: Penguin, 2004). Writes Moore, "A dark night of the soul is not extraordinary or rare.... To be sad, grieving, struggling, lost, or hopeless is part of natural human life."

Find out more about UNESCO World Heritage Sites at http://whc.unesco.org/en/list. These sites, designed to protect cultural and natural heritage from destruction, span the globe. But so far, the designation has had little to do with saving the natural night that has always been an integral part of these sites.

In the initial draft of his book (1947), Aldo Leopold originally placed his paragraph about "the penalties of an ecological education" in his introduction. He had lived long enough to see land that he loved ruined or damaged by his fellow humans and had suffered the anxiety, depression, and sadness that went along with such an experience. "I do not imply that this philosophy of land was always clear to me," he wrote. "It is rather the end-result of a life-journey, in the course of which I have felt sorrow, anger, puzzlement, or confusion over the inability of conservation to halt the juggernaut of land-abuse" (see *A Companion to A Sand County Almanac: Interpretive & Critical Essays*, edited by J. Baird Callicott [Madison: University of Wisconsin Press, 1987]). But, perhaps fearing such honesty would scare off otherwise sympathetic readers, he changed his mind and placed it instead in his essay "Round River," which appears on page 197 in most versions of the book.

Among his many accomplishments, Leopold was the driving force behind the creation of the world's first Federal Wilderness Area, the Gila, in New Mexico in 1924. While he was a writer all his life, his best-known work can be found in *A Sand County Almanac*, especially "Thinking Like a Mountain" and "The Land Ethic."

Roderick Nash's *Wilderness and the American Mind* (New Haven, CT: Yale University Press, 1967) is a detailed history of American attitudes toward our wild lands. Reading his account of the debates over such issues as the creation of our national parks or the establishment of the Endangered Species Act, one understands that contemporary debates—and especially attitudes and arguments that value the natural world as little more than a resource for monetary and material gain—have well-established historical precedents.

As Pierre Brunet of the Association Nationale pour la Protection du Ciel et l'Environnement Nocturnes told me: "I am pessimistic but I keep fighting. That is all. Why? I have to do that, it's my conscience. It's my duty to save the night environment. I appreciate stars, astronomy. There are many fighting for the environment. The night environment is a valuable fight. Nobody is fighting for that, so why not us?" Find out more about ANPCEN at www.anpcen.fr.

For more on solastalgia and Glenn Albrecht, the Australian philosopher who coined the term in 2004, see "Is There an Ecological Unconsciousness?," by Daniel B. Smith (*New York Times*, January 27, 2010). Looking for a term to describe the homesickness one feels while still at home, Albrecht created a word that has since found use around the world. "The growing influence of solastalgia is bittersweet for me," he told an interviewer in 2012. "As a philosopher you want your ideas and concepts to be influential and used and I'm pleased that people have found solastalgia to be inspirational, in the arts and in academia.... But equally, the concept in itself is depressing and it's unfortunate that people are all too familiar with the negative feelings it describes" (http://www.physorg.com/news/2012-02-solastalgia-bittersweet-success.html).

3: Come Together

The most recent addition to IDA's International Dark Sky Parks is Emmet County's Headlands property in Michigan, a six-hundred-acre stretch of forest on Lake Michigan just west of Mackinaw City. When I spoke with Dark Sky Park Program Director Mary Stewart Adams, she told me that the ultimate goal is a twenty-two-thousand-acre "dark sky coast" along the great lake. "Our ability to no longer see the night sky isn't just because of light pollution," Adams says. "We are losing our ability to dream and imagine." In response, Adams fills her programs with stories and myths, fairy tales and folktales from around the world, doing all she can to rekindle in her modern American audience an awareness about the night sky. She likes to ask, "If it's dark, then what?" As she explains, "every culture before ours has built temples and cathedrals and art, all of it asking, What is my relationship to the 'then what'? I'm hoping to inspire engagement with that question. I'm hoping to inspire the imagination."

One more story about Steve Owens and building community: Owens has designed a planetarium show for the blind. After working for several years at the Science Center in Glasgow, he came up with the idea of constructing a tactile hemisphere that would allow for blind people to "see" the night sky. Working with a colleague from the Design School in Glasgow, Owens placed pins in the patterns of constellations and then vacuum-sealed plastic over them to create a tactile night sky. Perhaps most ingenious was his and his partner's idea to simulate the Milky Way by using sawdust to evoke the countless spread of stars in the arms of our galaxy. Owens says that after months of testing he invited the press as well as four volunteers from the local charity for the blind and conducted a run-through of a half-hour planetarium show. "While I was doing it," he says, "I wasn't getting feedback from them, because obviously they were concentrating quite hard, and at the end, I said, 'So, was that okay?' and they just all said, 'That was amazing.' There was a woman there who was blind since birth, and she and her husband and kids go out to holiday cottages in the middle of nowhere and the kids are always talking about Orion and she said, 'Now I know what they're talking about. I could tell them how to find Orion. I could tell you what shape it is.' "

Owens says he and a friend, Adrian, recently conducted a meteor watch on Twitter. "I was in Glasgow, and Adrian lives down in Berkshire. I was having a barbeque; he was getting drunk, fiddling with his telescope. So we thought, 'Let's try to get a thousand people.' We got forty thousand the first night. It became the worldwide, top-trending topic on Twitter. The *Telegraph* had a headline, 'Perseids "Meteor Watch" Knocks Disney Star Miley Cyrus Off Twitter Top Spot.' The second night we were even busier, probably fifty thousand involved. Adrian and I were hosting it on Twitter. We had our laptops and were responding to questions. Almost everybody who was doing it was not astronomy-oriented in any way. People were saying, 'Is it safe to go outside? I'm not going to get hit by one?' So it's starting to be the case that even people who don't have an interest in science and astronomy are starting to hear about things. It's starting to soak into the public consciousness."

William Anders's photograph *Earthrise* has been called "the most influential environmental photograph ever taken." Of it, Anders has said, "There are basically two messages that came to me. One of them is that the planet is quite fragile. It reminded me of a Christmas tree ornament. But the other message to me, and I don't think this one has really sunk in yet, is that the Earth is really small. We're not the center of the universe; we're way out in left field on a tiny dust mote, but it is our home and we need to take care of it."

The time-lapse video of the Canary Islands everyone told me about is by Norwegian landscape photographer Terje Sørgjerd. It is entitled *The Mountain:* http://www.livescience.com/13739-mountain.html.

First constructed in 1671, the Paris Observatory (Observatoire de Paris) once sat outside the city under naturally dark skies. It now lies completely engulfed by the modern metropolis, no longer able to serve as an optical observatory but still quite a beautiful building. Even if you can't get in (it's not normally open to the public), simply wandering around imagining the lovely gray-white building set alone in a field, its astronomers observing the starry sky above, makes it well worth a visit. For an image of what it may have looked like, see http://en.wikipedia.org/wiki/File:Paris_Observatory_XVIII_century.png.

For a view of the Korean peninsula at night, search "Korea at night from space." The stark demarcation between the light-saturated South and the primitively dark North could hardly be more dramatic. Likewise, searching "world at night from space" will yield a selection of satellite photos showing not only the spread of electric lighting but also those places on Earth that remain primitively dark—primarily, places that either are economically undeveloped or without human inhabitants. So far, without exception, anywhere humans and economic development have moved, light pollution has followed.

Find the Museo Galileo online at www.museogalileo.it/en/visit.html. Find it in Florence housed in a typically beautiful Florentine brick building on the Arno River. In addition to its wealth of artifacts, the museum is also wonderfully uncrowded—there are moments when you can find yourself in the room alone with Galileo's telescopes, or surrounded only by four-hundred-year-old *globo celestes*. For an excellent recent article on Galileo, see "Galileo's Vision," by Davi Zax (*Smithsonian*, August 2009, 59–63), in which Zax accurately details the significance and drama of Galileo's story. See also *Galileo's Daughter*, by the always entertaining Dava Sobel (New York: Walker, 1999). After turning his homemade telescope toward the sky in 1609, Galileo wrote, "I give infinite thanks to God... who has been pleased to make me the first observer of marvelous things."

As is often the case with such topics, the question of who invented the telescope remains somewhat open to debate. Hans Lippershey, a German spectacle maker, is generally given the distinction as a result of his patent application from September 25, 1608. But others claimed to have come up with the idea before that. What is clear is that while Galileo did not invent the telescope, he built his own after hearing about the new creation and was the first (as far as we know) to use a telescope for astronomy.

Perhaps no one has done more to promote darkness in Europe than the indefatigable Friedel Pas. As the IDA's European representative, Pas has promoted dark sky efforts across the continent, and perhaps nowhere more effectively than in his home country of Belgium. Its annual Night of Darkness has grown to include two-thirds of the municipalities in the country, with some twenty-five thousand people directly involved. Pas points to the Night of Darkness as having an enormous impact on raising awareness about darkness, citing as a direct result the Flemish Parliament's having passed a unanimous resolution against light pollution just two months after the first Night. For anyone interested in combating light pollution, two things are key, Pas told me. First, "without awareness, you lose." And second, you must know more about the problem than anyone. You must, he says, be "weaponed with knowledge."

2: The Maps of Possibility

In the summer of 1898, forty-two-year-old John C. Van Dyke, accompanied by a fox terrier named Cappy, rode into the desert near what is today San

Bernardino. Over the next three years this art historian roamed the deserts of California, Arizona, and Mexico, and *The Desert* (1901; reprint, Layton, UT: Gibbs-Smith, 1980) is the result. His book is filled with highly attuned attention to desert detail, to colors and shapes and "the long overlooked commonplace things of nature," and to the feelings the desert's beauty brings. But an elegiac undertone runs throughout—Van Dyke knew that "every bird and beast and creeping thing" feared the arrival of humans because "they know his civilization means their destruction." We read his account from just over a century ago knowing he was right. "The fact that most of the beauty he described no longer exists," writes Richard Shelton in his 1980 introduction, "is too obvious to be dwelt upon."

Taking place originally on a beach near San Francisco in the late 1980s with a crowd of dozens, the Burning Man festival soon moved to Nevada's Black Rock Desert and now hosts crowds of forty-five thousand that come to revel in freedom and creativity for a few days at the end of every summer. The festivities culminate in a final burning of "the Man"—a wooden structure several dozen feet high on the festival's final night. Afterward, every trace of the festival—save for tire tracks etched into the flat playa—is erased. For more information, see burningman.com.

After landing on the moon, Buzz Aldrin said to a worldwide radio audience, "I'd like to take this opportunity to ask every person listening in, whoever and wherever they may be, to pause for a moment and contemplate the events of the past few hours, and to give thanks in his or her own way." He then ended radio communication and gave himself Communion, using a small kit given to him by his pastor. Aldrin has described his experience many times, including in a *Life* magazine interview (August 1969), and in his books *Return to Earth* (1973) and *Magnificent Desolation* (2009).

Isaac Azimov's short story "Nightfall" was inspired by Emerson's famous quote, "If the stars should come out one night in a thousand years how men would believe and adore," after Azimov's editor at *Astounding Science Fiction* countered, "I believe they would go mad."

Even if you can't read Italian, www.cielobuio.org is worth a visit for the photographs and simply to see evidence of all that this small group of volunteers is doing on behalf of darkness in northern Italy.

See the photo of Pisa, Italy, during Earth Hour at http://www.repubblica.it/ambiente/2011/03/28/foto/l_ora_della_terra_buio_sulla_torre_di_pisa-14176168/1.

Light-emitting diodes are leading the way in the transformation of lighting technology from electric lighting to electronic solid-state lighting. Far more efficient than electric light, highly programmable, and inherently directional (they only shine straight, rather than in all directions), LEDs offer the chance for us to address many of the challenges presented by too much electric light. Because of the heavy amount of blue light they cast, however, LEDs have raised concerns about human and environmental health, and manufacturers have yet to develop alternatives. Still, says the IDA's Bob Parks, "LED lighting has the potential to revolutionize outdoor lighting in a profoundly positive way."

In a sign that the French government is serious about cutting lighting at night in Paris to save energy (and money), a ban on lighting in and outside shops, offices, and public buildings between the hours of 1 a.m. and 7 a.m. will go into effect in July 2013.

Of the arguments on behalf of darkness, few have greater potential to effect change than the economic one—the fact that we waste so much money on lighting. Consider that "only 4% of the energy used to run a typical incandescent bulb produces light," writes Michael Grunwald in "Wasting Our Watts" (*Time*, January 12, 2009). "[T]he rest is frittered away as heat at the plant, over transmission lines or in the bulb itself, which is why you burn your fingers when you touch it." Grunwald builds a powerful case for efficiency as our greatest new source of energy, and argues one key change that must take place is for utilities to be able to save money through conservation and efficiency. In most states, utilities make more money by selling more power, and so have no incentive to do otherwise. Only six states have decoupled electricity profits from sales volume, Grunwald reports, but where they have, the results are impressive. In California and the Pacific Northwest, where utilities have actively promoted efficiency and conservation, "per-capita electricity use has been stable for three decades—while soaring 50% in the rest of the country."

While the lighting ordinances in Flagstaff and Tucson are the country's most well known, over the past ten years more than three hundred communities in the United States have adopted ordinances to control artificial light. Many are small towns, suburbs, or rural areas wanting to preserve the unique character of their community. In Florida, lighting ordinances have been used to protect sea turtles that nest on the beaches. For thousands of years, hatchlings have followed the light of the night sky to find the ocean, but hotels and streetlights have confused the turtles and drawn them inland toward their death. More than twenty-seven counties and fifty-eight municipalities have lighting ordinances to help. Elsewhere, the IDA and IESNA have teamed up to create the Model Lighting Ordinance (MLO), designed to make it easy for any community to adopt a lighting ordinance to control the use of light in their area. See www.darksky.org/MLO for more information.

Let There Be Night: Testimony on Behalf of the Dark (Reno: University of Nevada Press, 2008) features essays from twenty-nine writers, poets, and scientists, all of whom responded to a call to "speak a word for darkness" modeled on Thoreau's desire to "speak a word for nature" in his essay "Walking" (1863).

According to IDA's Pete Strasser, big box stores and other national retailers are often more eager to conform to a community's request for dimmer lights than are local contractors. "You know what Walmart and Home Depot do in Tucson? We tell them, 'We like these telescope-friendly dark skies; how about low-pressure sodium?' So the Walmarts here give us low pressure. Target has low-pressure sodium. They want to fit into the community. They don't have a standard. What holds us back is that the contractor putting in metal halide lights wants to sell a lot of poles, and wants to have a really juicy maintenance contract to replace those metal halide lights that burn out all the time. But if the municipality says, 'Hi, Walmart, will you please do this?,' they go, 'Sure.'"

Regarding Chris Luginbuhl's comments about how dimming the lights of a city like Chicago would have profound effects on the skies in surrounding suburbs and towns, Fabio Falchi told me something similar in Italy: "Because inside cities of course you cannot have a big improvement; the big improvement will be outside cities. Inside cities you can have an improvement in the comfort of the night, of not having light inside windows, in having lamps with colors that are not harmful for our health, so there are advantages also inside cities. But of the sky inside cities you cannot have a huge improvement. But outside, you can improve it a lot."

We are so used to the sky over our major cities being washed-out and void of all but the brightest stars that it can be nearly impossible to imagine what it would look like naturally, without all that light. The French writer Amédée Guillemin (1826–1893) created a series of small books on popular astronomy in which he included illustrations of the sky over Paris before electric light. His views of the Milky Way over the French capital (Paris: Le Ciel, 1866) are some of the most beautiful nocturnal scenes I know. See http://www.atlascoelestis.com/guil%2025.htm. Also impressive are the views of the night sky over London in 1869 by Edwin Dunkin from his book *The Midnight Sky: Notes on the Stars and Planets.* See http://www.atlascoelestis.com/22.htm. His book is available from Cambridge University Press (2010).

1: The Darkest Places

First published in 1968, Edward Abbey's *Desert Solitaire: A Season in the Wilderness* (1968; reprint, New York: Touchstone, 1990) remains the cantankerous and entertaining writer's most beloved work. Abbey would have enjoyed wonderful darkness during his time in Arches National Monument, just outside Moab, Utah. In the four-plus decades since Abbey's time in Arches, the National Monument has become a national park, but the town's light pollution has erased a significant portion of what Abbey would have seen.

Dan Duriscoe has been instrumental in the National Park Service's proposal for a "dark sky cooperative" covering the Colorado Plateau area in Utah, New Mexico, Arizona, and Nevada. The cooperative would attempt to preserve darkness in parks and communities in an area bordered roughly by I-40 to the south and I-15 to the west and north. The NPS will celebrate its one-hundred-year anniversary in 2016, and National Park Service director Jonathan Jarvis included the dark sky cooperative idea when outlining the NPS's ambitious goals for their next century of service. For Duriscoe, the Colorado Plateau idea flows naturally from a lifetime of advocacy for dark skies. Writing for *The George Wright Forum* (vol. 18, no. 4 [2001]), Duriscoe argued, in "Preserving Pristine Night Skies in National Parks and the Wilderness Ethic," that "the intent of the Wilderness Act of 1964 was to provide all Americans access to 'primitive and unconfined' recreation and opportunities for the spiritual enlightenment and personal development such experiences provide." If artificial light "compromises or interferes with the view of the night sky from a wilderness preserve," Duriscoe reasoned, "that light is in violation of one of the basic premises of the wilderness ethic."

On the clearest, darkest nights—no moon, no clouds—the human eye can still see, thanks to the night's natural light. While starlight contributes to this light to a small degree, it comes primarily from airglow, a weak glow caused by Earth's atmosphere that casts a uniform luminosity over the planet.

"The problem is that people don't recognize that there's a problem," says psychologist Peter Kahn, explaining "environmental generational amnesia." He argues that this concept "helps explain why we degrade and destroy the nature that we depend on for our physical and psychological well-being." This is certainly true when it comes to darkness and natural night skies—because most of us have no idea what we have lost, it doesn't occur to us what we are missing. A similar concept can be found in "diminishing baselines," the idea that each generation sees the world they inherit as normal, the baseline from which to judge change, even as that world is much diminished in natural beauty and wealth from the one their parents and grandparents knew. Hear Peter Kahn speak at http://histories.naturalhistorynetwork.org/conversations/environmental-generational-amnesia.

Thoreau's passage about contact (!) with the natural world comes from *The Maine Woods*, originally published in 1848. In his description of hiking Mt. Katahdin in Maine, he wrote, "Talk of mysteries!—Think of our life in nature,—daily to be shown matter, to come in contact with it,—rocks, trees, wind on our cheeks! the solid earth! the actual world! The common sense! Contact! Contact! Who are we? where are we?"

John Muir's words about our need for "beauty as well as bread" come from his book *The Yosemite*, originally published in 1912. A founder of the Sierra Club, Muir worked tirelessly on behalf of his beloved Sierra Nevada and was instrumental in gaining federal protection for Yosemite Valley. For a wonderful example of his exuberant writing style, see *My First Summer in the Sierra*, originally published in 1911.

The quote from Sigurd F. Olson about the value of "places where we can always sense the mystery of the unknown" comes from his *Reflections from the North Country* (New York: Knopf, 1976). Written near the end of his long life, this book is an excellent introduction to the longtime conservationist's thinking.

Despite the recent attention given by the National Park Service to preserving darkness, there is still much work to be done, admits Kevin Poe. "One of the things that's kind of frustrating is, even within the national park service community, this is not an issue that's on a lot of people's radars," he says. Thanks in large part to Poe, Bryce Canyon National Park offers more than 140 astronomy presentations a year, reaching more than thirty thousand visitors. But, by and large, whether an individual park pays attention to darkness hinges on the wishes of its superintendent. "The big goal would be to have darkness listed as a critical resource issue," Poe explains, "which would place it on a checklist for superintendents to address on a yearly basis. As it is now, in a lot of ways Chad [Moore], Dan [Duriscoe], myself, the team here, we're still these interesting characters crying in the wilderness."

In its "act to establish a National Park Service" in 1916, the U.S. Congress wrote, "The service thus established shall promote and regulate the use of the Federal areas known as national parks, monuments, and reservations hereinafter specified by such means and measures as conform to the fundamental purpose of the said parks, monuments, and reservations, which purpose is to conserve the scenery and the natural and historic objects and the wild life therein and to provide for the enjoyment of the same in such manner and by such means as will leave them unimpaired for the enjoyment of future generations." It is especially the goal of conserving the scenery and leaving it "unimpaired" for future generations to which dark sky advocates point. That the act also directs that the new secretary of the National Park Service receive an annual salary of $4,500, with an assistant director earning $2,500, one "chief clerk" at $2,000, one draftsman at $1,800, and "one messenger, at $600," has less contemporary interest.

That the boundaries of our protected natural areas are only as strong as we choose to keep them is a fact easily forgotten. All it takes for those boundaries to be breached is a severe enough economic downturn. Witness the state of Ohio's 2011 decision to allow natural gas drilling within its parks.

In addition to the work being done in national parks in the United States, parks in other parts of the world—most notably in the UK (http://www.nationalparks.gov.uk)—have been working to protect dark skies as well. Hortobágy National Park and Zselic National Landscape Protection Area in

Hungary, Izera Dark Sky Park in Poland, and Pic du Midi in France are just a few. For information on the UK efforts, see http://www.darkskiesawareness.org/dark-skies-uk.php. For information on the rest of Europe, start with the European office of the IDA: europe@darksky.org, or visit darksky.org.

Tyler Nordgren's *Stars Above, Earth Below: A Guide to Astronomy in the National Parks* (New York: Springer, 2010) blends astronomy and adventure as Nordgren spends a year traveling from park to park, sleeping outside as often as possible. A skilled photographer, artist, and writer, his book shares his love for the night sky and his valiant attempt to counter the fact that, "[s]adly, there are very few today who remember what the sky is supposed to look like at night, and those who do remember have grown used to the idea that it's just the way things are now." Nordgren's artwork on behalf of the U.S. national parks draws inspiration from the style depicted on posters produced for the parks between 1938 and 1941 as part of the Works Progress Administration (WPA) Federal Art Project. Initially funded in 1935 and lasting eight years, the WPA helped put millions of Americans to work during the Great Depression.

Like many other nineteenth-century American writers, Walt Whitman showed an intimate attention to darkness and night in his work. His famous *Leaves of Grass* (from which "The Learn'd Astronomer" [1900] comes) is full of night imagery, and poems such as "Out of the Cradle Endlessly Rocking" feature the poet experiencing firsthand the world after dark. One reads his exhilarating work and wonders if a contemporary poet would ever think to rhapsodize about darkness in a similar way.

Actually, according to the 2010 census, the mean population center of the United States is Texas County, Missouri, but Chad Moore is close.

In *One Square Inch of Silence* (New York: Free Press, 2008), Gordon Hempton goes looking for and finds exactly that in Washington's Olympic National Park.

William Fox's books include *The Void, the Grid & the Sign: Traversing the Great Basin* (Reno: University of Nevada Press, 2005) and *Mapping the Empty: Eight Artists and Nevada* (Reno: University of Nevada Press, 1999). He is director of the Center for Art and Environment at the Nevada Museum of Art in Reno. James Terrell's Roden Crater sits at the western edge of the Painted Desert in the San Francisco Peaks volcanic field outside Flagstaff, Arizona. It has yet to open to the public.

Index

Note: *Italic page numbers refer to figures.*